SEVEN DEADLY SINS

Also by Guy Leschziner

The Man Who Tasted Words:
A Neurologist Explores the Strange and
Startling World of Our Senses

The Nocturnal Brain:
Nightmares, Neuroscience, and the
Secret World of Sleep

Seven Deadly Sins

The Biology Of Being Human

Guy Leschziner

ST. MARTIN'S PRESS
NEW YORK

First published in the United States by St. Martin's Press,
an imprint of St. Martin's Publishing Group

www.stmartins.com

The Library of Congress Cataloging-in-Publication Data is available upon request.

ISBN 978-1-250-28881-3 (hardcover)
ISBN 978-1-250-28882-0 (ebook)

Figures by Martin Brown

Our books may be purchased in bulk for promotional, educational, or business use.
Please contact your local bookseller or the Macmillan Corporate and Premium
Sales Department at 1-800-221-7945, extension 5442, or by email at
MacmillanSpecialMarkets@macmillan.com.

Originally published in Great Britain by William Collins,
an imprint of HarperCollins*Publishers*

First U.S. Edition: 2024

10 9 8 7 6 5 4 3 2 1

For Kavita

Contents

Introduction

Out of the crooked timber of humanity, no straight thing was ever made.

Immanuel Kant

A journey through the domains of sex, murder, infidelity, criminality and violence resides within the pages of this book. These facets of the human condition do not exist in isolation though. They are woven through with the robust threads of genetics, neuroscience, evolutionary psychology and pathology. It is these aspects of our biology that are the ingredients that define those human forces that have indelibly shaped our world, the primal constituents of these acts and other transgressions.

The sins – those emotions and deeds decreed as the origins of all wrongdoing by religions, theologians and philosophers – have built and destroyed empires, fuelled the expansion of humankind into every corner of the world, and even beyond it. They are the architects of human ascent and obliteration. They have driven the amassing of huge fortunes, the search for resources, wars that have gone on for generations. The ebb and flow of human history is defined by the Seven Deadly Sins: wrath, gluttony, lust, envy, sloth, greed and pride. From the wrath that has ignited revolutions, to the greed that has re-sculpted the world map. From the sloth that has led to the fall of empires, to the envy that has built them. From the lust that has led to the fall of politicians and the betrayal of national secrets, to the voracious gluttony that has left our environment in ruination, and the pride that has fuelled countless conflicts.

Our less savoury tendencies are not only the engines of global history, however. They are the forces that fashion our present. Our national borders, our politics, our economies, the fundamental nature of our societies. In my clinics, I see the consequences daily – Syrians, Afghans, Iraqis, people from former Yugoslavia – all in London as an outcome of war and upheaval, the fruits of the human sins. And as I look at the streets around the hospital, the myriad faces is testimony to other cruelties inflicted by man. Colonisation, slavery, trafficking, conflict – products of greed, wrath, envy and pride. These sins are the drivers for the personal lives and stories of every single one of these individuals, and indeed for us all.

The last twenty-five years of my life have been a window into human society, delving into corners of humanity that relatively few outside the world of medicine ever see. Disease is a great leveller, afflicting every manner of person, striking without prejudice, from the vicar to the murderer. It is one of the greatest privileges that medicine affords: insights into every patient's life, no matter who they are or where they come from. Life in a hospital exposes one to the full spectrum of human morality. Inexplicable altruism, generosity, kindness and love. The woman who donates a kidney to a complete stranger; the kind Samaritan diving into the murky waters of the Thames to save a drowning man. The pure goodness of those in our society who care for others, who go well beyond what is required to earn a living. But also, unspeakable cruelty, sloth and gluttony. Patients eating themselves to death, or those people brought into the emergency department beaten to oblivion, the victims of anger, envy, pride, lust.

These human failings have also affected my own life. If I examine where I am now, what I am doing with my life, my own thinking, my own psychology and world view, I can directly trace them back to the sins, those foundations of wickedness.

Since I learnt to talk, I have had to correct the pronunciation of my name (although I have largely given up now). The maelstrom of consonants, the alien juxtaposition of S, C, H and Z, strikes panic in the hearts of dentists' receptionists calling out my name, colleagues introducing me, patients, even cold callers asking if I have been involved in an accident. And it is not just native English speakers. Even Germans, who should be a little more familiar with the sounds, and the phonetic spelling of my family name, approach it warily.

The reason for this unfamiliar name is its unusual roots. The family originates from an area of Europe called Silesia, itself victim to countless conflicts, territorial aggression and greed. Now within the borders of Poland, this region has over the last millennium been under the control of the Silesian Piasts, the Mongols, the Bohemians, the Hungarians, the Polish kings, and then the Prussians, becoming part of the German Empire in 1871. After the First World War, following a plebiscite and a subsequent uprising by ethnic Poles unhappy with the result, Silesia was divided between Germany and Poland. My grandfather, born in the pre-plebiscite German city of Breslau, remained in the post-division German part of Silesia. The origins of the name, however, are probably from a more Polish bit of Silesia, from a small village called Leszczyny, although the name has been bastardised, spelled using phonetic German, with the -er suffix denoting 'from', much in the same way as Berliner or Frankfurter. Hence a Polish name, germanicised, sowing seeds of confusion everywhere in the world.

My grandfather – imprisoned on Kristallnacht in 1938, and after a short period in the Buchenwald concentration camp – was extraordinarily lucky. He and his brother were released through a little-known and little-recorded rescue scheme for 4,000 adult men imprisoned in Sachsenhausen, Dachau and his own camp. Between February 1939 and the outbreak of the Second World War in September 1939, these 4,000 men, my grandfather and his brother included, left behind their families (who all subsequently perished in the camps), boarding ferries from Ostend to Dover. A short while after being housed in Kitchener Camp – an old First World War base on the outskirts of Sandwich in Kent – he enlisted as an aircraft engineer in the Royal Air Force, and so began twenty years of a very peripatetic existence, encompassing Europe, North Africa and the Middle East. He settled in Switzerland in the late 1960s, seeking tranquillity and security.

Thus, my inheritance is also the product of these human sins. The envy of others, the greed for natural resources (the *Lebensraum* the Nazis so craved), the naked, cold aggression, the pride or hubris of fascist dictators – 'man's inhumanity to man'. On my mother's side too, centuries of life in Baghdad, uprooted in hatred and violence; leaving, then cut adrift as refugees. For my paternal grandfather, first-hand witness to the horrors of fascism emboldened, he saw his parents, uncles, aunts and cousins, every member of the wider family exterminated, by people they considered their compatriots. This is the ultimate reason for the rarity of my family name. This stain of evil clung to my grandfather until his dying day, his faith in mankind shattered beyond redemption, his life narrowed to his wife and family.

I do not think that my parents' generation, or indeed my own, have entirely shaken off the shadow of my grandfather's trauma, a supreme expression of human sin. It lives on in a barely suppressed pessimism, a latent fear or anxiety about what might happen, what the future holds. Many people have written of the immigrant mentality of having a suitcase packed. Just in case. In preparation for the time when history does indeed repeat itself, when the ugliness of human beings arises again. And I felt this strongly as a child. Some of this was said, snippets of conversations with parents or grandparents, but much of it was no less understood in the absence of words.

I realise that I am far from alone in having this sort of story of human cruelty, inhumanity, of evil, as my inheritance. The Holocaust is only exceptional in its industrialisation of sin, but wholesale slaughter – genocide – is not exclusive to the Second World War. It was happening long before and is still happening – from Carthage in the Third Punic War in 149 BCE, to various places around the world today.

In the examination of the 'sins' that follows, I seek not to necessarily excuse these and other tragedies, to explain the apparently inexplicable. Perhaps, however, a starting point is to understand the building blocks of the human experience, the essential nature of the emotions and deeds that, at their most intense, are the germs of these global and personal events, but also many other facets of our lives. To at least grasp the components that at their extremes give rise to these aspects of our existence. To gain insights into why we do what we do: the biology of being human.

* * *

The Seven Deadly Sins were immortalised in the public imagination by Dante Alighieri in his fourteenth-century poem, *The Divine Comedy*. As Dante ascends from Hell, he discovers the Mountain of Purgatory, terraced into seven levels, each one representative of one of the seven roots of sinfulness. Pride resides on the first terrace, the souls of the proud doubled over by the weight of huge rocks on their backs. The second terrace is envy, then wrath, sloth, greed and gluttony. At each level, the souls of sinners endure punishments befitting their sin. At the highest terrace, one below the peak of Earthly Paradise, souls must jump through a burning wall of flame, crying out examples of lust.*

However, the theological origins of the Seven Deadly Sins are much more ancient. Dante relied upon the Christian tradition of his time, which in turn leant on Judaic theology. The Old Testament viewed sin as the violation of God's commandments, both in behaviour and thought. Talmudic rabbinical literature was increasingly concerned with how thoughts and feelings relate to our actions, and might conspire to disobey God. These Jewish principles were formalised by the Desert Fathers, early Christian hermit monks residing in the Scetes desert of Egypt, in the fourth century, and were listed as eight sins. It was Pope Gregory I in 590 CE who revised the deadly sins into the more familiar Seven Deadly Sins format – lust, gluttony, greed, sloth, wrath, envy and pride. Saint Thomas Aquinas described them as 'capital sins', the foundations of all other sins. These immoral thoughts and actions were the basis of all our contraventions of the laws of God and man.

It is not just the Judeo-Christian world that has been preoccupied with sin, and its classification and categorisation, however. So too have all the other world religions, and indeed theologies now extinct, such as Greek and Roman mythologies. These human failings also fascinated the ancient philosophers, from Plato and his tripartite view of the soul – reason, desire and emotion, all competing to influence our behaviour – to the Stoics, who proposed the abandonment of worldly pursuits in the search for freedom and happiness.

That this theological and philosophical concern with sin is

---

*   In the poem, two groups of sinners move in different directions, Sodomites in one, the Pasiphae in the other. Subsequent illustrations of Dante's Purgatorio often omitted this particular aspect, as it implied that Dante viewed homosexuality as no particular obstacle to salvation.

so universal, transcending time, geography and culture, simply reflects that these thoughts and acts are ubiquitous too. These sins are hard-wired within us, deeply embedded in the recesses of our brains or the essence of our souls. And that they have had such profound influence over human history, and indeed the organisational structures of our societies today, hints at their dualistic nature – that these 'sins' may have benefits as well. For if these behaviours were solely agents of harm, why should they be the weft of the tapestry of who we are? What possible reason would there be for such destructive traits to be passed down through the generations, to persist throughout the evolution of life as we know it?

* * *

My views on human nature have undoubtedly been guided by my clinical practice – by meeting patients with these traits, not caused by inherent moral weakness or evil, but the consequence of disease or injury, where abnormal function of the body results in a medical disorder. It is this that has left me with a more nuanced perspective of human 'failings', the origins of our 'bad behaviour', of inadequacies, of the frailties that constitute who we are. It is all too easy to ascribe a moral value to some individuals, to perceive them as sinful, immoral or weak. However, as the people in this book will demonstrate, this unambiguous view is clearly overly simplistic. Disorders of the brain, of our genes, or other physical conditions, may give rise to gluttony, lust, wrath or pride. The effects of our environment or our upbringing may produce envy, lust or sloth. Crucially, these disorders unmask what is already in us, what already exists in all of us.

If changes in our biology or psychology can give rise to these emotions or actions, then this implies that they derive from our physical make-up, the configurations of our bodies and minds, rather than our 'souls'. That all of us have the propensity to 'sin'.

These human characteristics sit buried within us all, defined by our genes and by our evolution, moulded by our environment. As we will see, their omnipresence implies that these traits are fundamental to our survival and success, and to consider them flaws is not entirely correct: they serve evolutionary imperatives to save us, to preserve the tribe, to ensure the advancement of our societies. These aspects of our character may cause terrible cruelty and suffering, but can also serve a useful purpose, a potent driver of the triumph of our species. And while they do not excuse the worst of our natures, nor ease my understanding of my own family story, we cannot ignore the essential foundations of our transgressions, and how they shape our history, our present and our future.

For millennia, the basis of human wrongdoing has been framed by theologians and philosophers in moral terms, as sins, transgressions against divine law: that all our shameful, selfish acts are moral failings, all originating from the Seven Deadly Sins. That it is these moral flaws that form the foundation for all evil in the world, offences that slight God and humankind.

Maybe, however, it is time to reconsider all this talk of sin in light of our increasingly secular twenty-first century world. For all of us, these 'sinful' character traits are perhaps less of a moral issue and more of a biological one, raising questions of responsibility, blame and free will in the face of sin. It is only at their extremes that they give rise to untethered human suffering, pain and tragedy. The question is whether the essence of these emotions and behaviours truly represents sin, or simply reflects the unbridled intensity of our intrinsic drive to survive and thrive. And where the boundaries between normal human nature, human pathology and sin are actually drawn.

1.

Wrath

*Some through too much passion have burst their veins . . .
and sickly people have fallen back into illnesses . . . many
have continued in the frenzy of anger, and have never
recovered the reason that has been unseated . . . it conquers
the most ardent love, and so in anger men have stabbed
the bodies they loved, and have lain in the arms of those
whom they have slain.*

Seneca, *On Anger*

I see Sean* increasingly rarely these days, as he prefers to talk
on the telephone or via video conferencing. The journey to the
hospital is perilous for him, and he fears trips on the London
buses.

I vividly remember the first time I met him. I had his old
medical notes to hand, and could see references to anger, aggression and arrests alongside his long history of epilepsy. When I
first called out to the waiting room, and I saw him stand up, I
was already filled with trepidation. Then in his late sixties, Sean
was huge, several inches taller than me, with arms thicker than
my thighs and a bull-like neck. As he strode into the room, I
could see a slight trembling, and his face was reddened. His

---

* Name and some details changed.

entire body was taut with energy, like a bomb about to go off. This did nothing for my nerves, and as he sat down in a clinic chair, I kept my distance, making sure that I sat closer to the door than he did. My eyes were drawn to the veins in his neck standing proud, his fists clenched tight.

As we discussed his symptoms, I with one eye on my exit, a story evolved of someone who had had seizures for many years, that had plagued his life since his twenties. Regular seizures, two or three times a week, uncontrolled by any of the several drugs he had been treated with. Not full-blown convulsions, with resulting shaking, loss of consciousness and collapse to the floor, but no less distressing. As I learnt over the following months and years, what I had taken to be the body language of anger and aggression was actually the manifestation of deep anxiety, the bone-crushing constant fear of a seizure at any moment, compounded by a dark pit of depression intimately linked to the effects of the condition on his life.

While his seizures were in some ways less dramatic, limited to a small part of his brain rather than enveloping the whole of it, in other ways they were even more overwhelming. From his perspective, he would only get a few seconds' warning, a strange feeling in his stomach, a delicate churn, a rising sensation like dropping down in an elevator, before he would then lose awareness. In his confusion after the seizure, he would be aggressive and violent, lashing out with abandon. He would come to, unaware of how long had passed or what had happened, memories of his own life in tatters. In the aftermath of these seizures, he would sometimes find himself surrounded by destruction, furniture in pieces, smashed glass, occasionally in handcuffs, pinned down by police officers. Arrested, charged, detained, sporadically sectioned in a hospital room. He lived in fear of these seizures, and the damage he could wreak; a hermit venturing rarely outside the confines of his home, imprisoned by his neurological disorder. Hence his reluctance to come to hospital: his trips for clinic appointments had ended up in a police cell on more than one occasion. For in the wake of his brief seizures he would 'rage like a wild beast', filled with anger and violence, but without any recollection at all. His body and mind, entirely invaded by another being, a dark force, for a few minutes at a time, without warning or obvious triggers. The unpredictability of his seizures, and their impact on how he had been treated and perceived, had diminished his life as much as the diagnosis itself.

This perceived relationship between epilepsy and violence has been noted before. The nineteenth-century Italian physician and criminologist Cesare Lombroso contributed greatly to the stigmatisation of people with epilepsy. His work in forensic medicine culminated in his magnum opus, *L'uomo delinquente* ('The Delinquent Man'), in 1878. His theory of criminality proposed that criminals differed from non-criminals physically as well as mentally, and represented a reversion to primitivity, a form of reverse evolution suiting pre-civilised human society. He also strongly linked epilepsy with criminality and violence. His views were not particularly groundbreaking. The belief that epilepsy was linked with violence and aggression had been around since ancient times, but he did much to promote these ideas.

In truth, however, aggression due to epilepsy is rather rare indeed. Almost invariably, it reflects neurological dysfunction – a disordered brain directly attributable to seizures – rather than an underlying predisposition to violence.[1] Organised aggressive behaviour, directed at something or someone, during a seizure is very unlikely. While seizures themselves may trigger intense uncontrolled electrical activity in areas of the brain precipitating fear, or areas that encode behavioural responses to threat,[2] these seizure manifestations are usually brief, and unfocussed.

For most people like Sean, the aggression stems not directly from within the seizure itself, but from its after-effects. Seizures arising in and limited to one part of the brain may not simply dissipate. They may leave the brain, as a whole, affected. The ripples of that seizure may reverberate within the electrical circuitry for minutes, hours or even days afterwards – the so-called post-ictal state. Seizures cause changes in chemistry, blood flow or inflammation, that can prevent normal brain function for quite some time afterwards.[3] What is left behind is a brain disrupted, yet to recover from an electrical storm.

In many cases, seizures may give rise to confusion in their aftermath, with aggression arising from the disorientation. In one study from Japan the authors described a thirty-one-year-old woman who had had epilepsy since the age of two. She was known to have seizures arising from the temporal lobe that had failed to be brought under control with medication. She had been in a souvenir shop in a tourist site, when she suddenly began fumbling with the goods on display, indiscriminately removing them off the shelves. The shop staff, alarmed by her actions, tried to intervene, initially by shouting at her, but she did not respond. Increasingly panicked by the disruption to the shop, one of the staff tried to restrain her physically, but he was knocked to the ground, and the police were called. She continued to rummage along the shelves, completely unaware of the fuss around her, until the police arrived. It finally took six police officers to wrestle her out of the shop and get her to hospital. By the time she was reviewed by doctors, she was completely back to normal, but without any memory at all of the incident.

In other cases, however, these sorts of seizures may even give rise to psychosis. Between seizures these individuals may be totally normal, but after one, or sometimes a cluster of them, there may be a period of delusions, hallucinations* and mood disturbance, indistinguishable from someone psychotic as a result of psychiatric illness. The only clue may be the periodic nature of the psychosis and its temporal relationship with seizures. One man in his forties, leading an otherwise normal and healthy life, appears at our hospital roughly once every year or two after a bout of seizures, and requires several burly mental health nurses to stop him ripping sinks off walls and assaulting other nursing staff. He remains under section for a week or two, before he returns to life outside the hospital. The transformation in him is astounding – two separate minds in one body, switching with the blink of an eye, or more precisely a spark in the brain.

The good and bad of anger

In cases like Sean's, where wrath is a product of disease or disorder, there is a clear abnormality of brain function that gives rise to it. Yet all of us have a propensity to anger; wrath is not just the product of an abnormal brain. It is, after all, a natural emotion, part and parcel of being human. Understanding normal anger is crucial to our knowledge of pathological anger, and vice versa.

Among our negative emotions, anger is unusual. Unlike sadness, fear and disgust, which lead us away from the provocation, anger drives us towards it. To confront, to fight.

When uncontrolled and unfiltered, anger can be destructive,

---

* Delusions are defined as firmly held false beliefs despite very clear incontrovertible evidence that disproves them. Hallucinations represent the hearing, seeing or otherwise sensing of something that does not exist. These are the features of psychosis, but hallucinations in particular can occur in situations outside psychosis.

either through the intensity of the emotion, as in rage, or in its physical manifestation of violence. Anger has some very clear positive aspects, however. If individuals are given a puzzle to solve (one that is actually unsolvable), some respond with despair or dejection, while others respond with anger. When given a second puzzle, this time one that does actually have a solution, those who experienced anger with the first perform much better, and persist in their efforts for longer.[4] Anger is the emotional response to being thwarted, being treated unfairly or not receiving an expected reward, when situations or the actions of others block our ability to achieve a goal. Anger is a motivator, a drive to continue striving to reach one's objective.

But as a driver of behaviour to achieve, to attain, to possess, it can sometimes lead to aggression. Most people experience anger very regularly without recourse to violence, though not all. While anger is the emotion, aggression is the behaviour – intended to harm, to inflict damage on someone else. And while anger does not automatically lead to violence, high levels of anger as a personality trait predict aggression, and domestic violence, poorer functioning in the wider world and interpersonal problems.

Thus, while the origins of wrath are anger, it is the product of it – aggression and violence – that is most feared. It is the nature of this relationship, between the emotion and its resulting behaviour, that is most concerning. Why is it that some people are quicker to anger, and swifter in their response to it?

Even this model of cause and effect – of anger leading to violence – is overly crude, since aggression comes in varying forms. There is reactive aggression: violence that is impulsive or defensive, spontaneous, emotion-laden, with the purpose of dealing with a threat in the heat of the moment. Then there is proactive aggression, premeditated, for a broader purpose; colder. This latter form may of course arise from anger too, but anger is not a prerequisite. We humans are particularly predisposed to this form of aggression, more of which later.

Iatrogenic rage

While cases like Sean's – where anger and aggression are the aftermath of a seizure – are very rare, there is another clinical scenario that is unfortunately much more common. The most frequent cause in people with epilepsy is probably doctors – myself included – where rage is 'iatrogenic': illness caused by medical examination or treatment.

Jono could be characterised as a 'gentle giant'. He is imposing, tall and broad-shouldered, but his physical presence is completely disarmed by a very ready laugh, and a frequent joking self-deprecation. Now twenty-nine, he trained first as a primary school teacher, then as an accountant – neither occupation known for their violent nature. He is now head of finance for a law office in the southwest of England. Hannah, his wife of a year, is a self-confessed introvert: 'Jono is an extrovert, a social butterfly, the life and soul of the room. So we make for a good fit.' She works as a criminal lawyer, and they met a few years ago as colleagues in another law firm. She tells me that one of the things that she found so attractive about Jono was his completely unflappable nature. 'He was this really easy-going guy. Nothing fazed him. He was a very mellow individual.'[5]

On their second date, Jono told Hannah of his diagnosis of epilepsy. He had had his first convulsion the day after his eighteenth birthday, rolling home at 4 a.m. after 'a few too many lager-shandies'. Initially, doctors ascribed his collapse to a faint, but after two further convulsions over the next few weeks, and some tests, a firm diagnosis of epilepsy was made. During an

EEG – the application of electrodes to the scalp to record the electrical activity of his brain – he had briefly become unresponsive and had lost of few seconds of awareness, a fleeting tuning out from the world. Obvious epileptic activity was visible on the brainwave traces, the hallmarks of an 'absence' seizure: widespread disruption of electrical signals causing his brain to stutter for a brief moment, before resuming normal function. In hindsight, Jono thinks that he had been having absences for quite some time. 'I always thought there was something not quite right. I would be having a conversation, and I'd lose track. I would be off in the distance. I would shudder, and suddenly be right back in the room. It would even happen when I was talking. I would forget what I was saying, and people would look at me, wondering if I was still there.'

When Jono and Hannah met, his epilepsy was well controlled. He had been started on an anti-epileptic drug immediately after his diagnosis. Despite leading a relatively normal life at university – albeit with some caution about sleep deprivation and alcohol, both potential provocation for seizures – his epilepsy had been quiescent. An unfortunate exception to this had been a seizure that had occurred while teaching a class of primary school children as a student teacher. His convulsion in public view was traumatic for both the children and Jono. 'We agreed with my university that I was probably not in a safe way, being left with small children on my own.'

He completed his teaching degree, but this incident led him to seek accountancy qualifications. By the time he met Hannah though, the epilepsy had melted into the background, and Hannah did not dwell on his diagnosis. 'There was nothing affecting him, as far as I was aware,' she tells me. 'I learnt about the not drinking [a lot], and not being out too late.' While she realised that occasionally Jono would not want to go out due to being tired, not wanting to risk another seizure, otherwise it left their relationship unaffected. 'I was a pretty good boy at that time,' Jono laughs. 'I lived my early twenties like a normal twentyish-year-old lad, just with a few limitations on life's indulgences. I managed to get my driving licence back. But then it all changed, quite suddenly out of the blue.'

A year into their relationship, his seizures came back with a vengeance. Not having had any for several years, without apparent reason he began to have convulsions. 'I was working at a new company, really enjoying it. And I just decided to have a seizure one day. It was a lunchtime, during the working day. I just decided to do it in the office, just to show everyone that, hey, he is disabled,' he chuckles. I ask him why he chooses the term 'decided to', and Jono becomes more sober. 'It's a coping mechanism, a touch of humour. That it is me, not that dark bit of me.' An attempt to wrestle back a little bit of control over his life – a control that the epilepsy has snatched away.

After the seizure in the office, Jono's convulsions came thick and fast, one every six to eight weeks. It is the nature of epilepsy to be unpredictable. Sometimes these dramatic resurgences have a clear cause – sleep deprivation, missing medication, illness, for example – but often no obvious reason can be found.

Hannah's first witnessing of his seizures is etched into her memory. 'He had come over to my house after work, and he was lying on the sofa. I was cooking dinner. I was talking to him, and then he wasn't responding to me,' she tells me. 'I poked my head round the door, and he was having a seizure on the sofa. His lips were blue, and he was making a sound like he couldn't breathe. Eyes rolled back, with his entire body going. Rhythmic jerking. Instinct just kicked in.'

Since then, Hannah has observed many more of Jono's seizures, each rather similar. Inside the house, on the driveway, on one occasion at his cousin's wedding. 'That was the worst one,' Jono says. 'Hannah and I were dancing together, holding hands, proper romantic. I decided to have a seizure on the dance floor, in front of the entire family. I still feel so sorry for doing it. Whenever I see my cousin, I still feel so guilty.' Again, the 'decided to'.

The seizures are often preceded by head-turning, then the convulsion, which can last up to five minutes. Afterwards, he remains still for a short while, his breathing gurgling, almost as if he is underwater. Occasionally, there will be blood, when he has injured himself. Then a sudden start – 'as if someone has poked him with a stick. But then he doesn't recognise who he is talking to, does not remember anything that has happened. It is quite traumatic,' says Hannah.

Jono's dosage of medication had already been pushed up as high as possible. To re-establish control, Jono's specialist prescribed another anti-epileptic drug at a high dose, a drug called levetiracetam (frequently referred to by its original trade name, Keppra, for ease). It is often a very effective medication, and unlike some of the older medicines we use, without inter-actions and generally very safe. It is a drug that I prescribe very often indeed. This new drug was transformative for Jono, rendering him seizure-free almost immediately. He has remained without seizures for four years now.

The drug has been transformative in other ways too, however. 'I was told that I would experience mood swings for a couple of weeks while I was getting used to it,' Jono tells me. But this was not entirely accurate. Four years on from starting on it, he says he feels like he has 'a demon within him'. While he has indeed had mood swings, these have not settled. And the worst of it is something that is sometimes described as 'Keppra rage'. As Hannah has noted, Jono's underlying character is relaxed, calm, unruffled in the face of stress or aggravation. Now, however, the drug has done something to him, to the emotional regions of his brain. 'I always feel like my baseline is off,' he says. 'I'm always waiting for the next time that I'm going to blow my top off. I just feel like I'm on idle, getting ready to be quite vicious in the way I talk to people.'

There are countless examples of him letting rip: episodes of road rage, swearing at a woman blocking the aisle at their supermarket, lashing out at family or friends. At work, he has sworn at his colleagues – in one episode humiliating his boss, cursing and belittling him in front of the entire office. Later, his boss, also a friend, took him aside. 'I'm talking to you not as your boss now; I'm talking to you as a friend. Yes, something's not right in your head at the minute and there's something external causing you to not be the same person that you were. But you can't go acting like this, you can't think that there's going to be no consequences to you being like this around other people.' His boss feared that Jono's outbursts would ultimately result in him being punched or subjected to other acts of retribution.

In the aftermath of these events, when the anger has subsided, Jono is deeply apologetic, realising he has crossed a line. In the heat of the moment, however, he is unable to contain himself. Hannah says: 'All of a sudden, this angry person, completely different. It is never directed at a single person. It is just anybody, everybody.'

It is not just verbal aggression. He has smashed furniture in a fit of anger. He also tells me of a baseball game he was playing. It was his turn to bat: the opposing team 'were quite lippy anyway, so I was already feeling a bit buzzed. And their pitcher, either accidentally or not, decides to throw the ball at my head. I was there with a baseball bat. So I started running at him. Hannah came running out, and my teammates had to take the bat off me, to prevent me smashing in his head. I was going to cave it in.'

Jono played competitive rugby for many years and had never

even had a booking for misconduct or dangerous play. He had never had a fight. 'I had never had this side to me. It wasn't until I was on Keppra that it just started to happen. In those moments, I feel like I'm not at the wheel. This angry dragon is taking control. And it's got all the adrenaline in the world to break bones and spit fire. It's not me,' he says. 'He often describes it as watching himself, when he calms down,' Hannah adds. 'He is not in himself, but watching from the outside.'

It is difficult to put oneself in Jono's shoes. As we talk, I can see the levity in him, the calmness and unflappability that drew Hannah to him in the first place. His intrinsic character is diametrically opposed to the wild displays of anger and aggression they describe. To know that this loss of control, this rage within him, can arise like the flick of a switch, must be disconcerting, if not frankly terrifying.

Levetiracetam, the drug Jono was prescribed, is far from the only anti-epileptic drug that can cause behavioural disturbance, sometimes even psychosis.[6] These drugs, after all, alter brain biochemistry and the transmission of electrical signals in the brain. Its widespread use, the dramatic effects it rarely has, and its usage for many years, perhaps unfairly tar this particular drug with the rage that most practising neurologists will have seen at some point in their own patients. One of the first patients I saw as a young neurology registrar was a frail elderly lady, her grey hair immaculately curled, who had been prescribed this drug. She was forcibly admitted to hospital after being tackled to the ground by several police officers in her front garden. Its ability to control seizures is why I have not lessened my prescribing of it, despite this and other experiences. However, I am certainly cautious to warn people of the possibility of these effects, and that alternatives are available.*

---

* The nature of levetiracetam's effects on behaviour are fully not understood,

Hannah says the violent edge to Jono's anger has eased in the last couple of years, though he remains prone to outbursts. The triggers remain the same. A feeling of being humiliated, pride being dented, an element of threat – either to him or to Hannah; sometimes, it might be just mild frustration or a minor irritation. Despite them having a good relationship, not prone to arguments, occasionally his explosive outbursts can cause trouble between them. Although Hannah says she has never felt at risk and is sure he would never be violent towards her, she feels sometimes that it is like walking on eggshells, not sure if something at work might have angered him, and might spill over into their domestic life.

I am surprised that Jono has never actually raised his rage issues with his neurology team. The letters detailing his appointments report that he describes some temper issues, but it is clear he has downplayed them, and their effects on his and Hannah's lives. Perhaps he is so grateful that his seizures have abated that he feels he has no option but to remain on this drug for the foreseeable future. This is not necessarily the case, and Jono does have alternatives. He is not destined to live like this for the rest of his life.

Hannah is circumspect. 'Jono is incredibly intelligent, incredibly funny. All the good qualities of a human being, Jono has got. When he is angry, when he is attacked, when something is

---

but it has a cousin, brivaracetam, a relatively new drug that is structurally very similar. The younger pharmacological cousin is much less likely to cause these behavioural changes, and researchers have pointed out that levetiracetam has an effect on a particular neurotransmitter receptor in the brain, the NMDA receptor, that the newer drug does not.[7] This receptor detects a chemical called glutamate, and is thought to be fundamental to learning, memory and neuroplasticity. In addition, however, blocking these receptors, through drugs or occasionally through auto-immune disease – where antibodies directly binding to these receptors are produced within the central nervous system – will often cause marked behavioural change, sometimes even psychosis. The presumption is that this NMDA receptor is a crucial component of the circuitry that governs the generation of anger or its inhibition.

happening to him or those around him, all those nice qualities disappear. It is like a completely new person. But he comes out of it pretty quickly. He is aware when he has overstepped the mark.' Both she and Jono describe two people living in one body: a Jekyll and Hyde. The 'monster' is a creation of the pill he swallows morning and night. But that monster will also die with the cessation of that pill.

Neural roots of anger

What Jono and Sean clearly illustrate is that alterations in brain activity or brain chemistry can act as the amplifier of that very normal emotion of anger, and resultant aggression. And by implication, that these emotions and acts emanate from the brain. What then do we know about the neural origins of anger, outside the constraints of the neurology clinic?

From a psychological viewpoint, some of this is related to a very particular type of attention. People who are generally more angry pay much more attention to stimuli in their environment that are associated with hostility. They are more likely to pick up on perceived hostility or potentially anger-inducing stimuli than others.[8] Whereas for most of us the scowl of a passerby may go unnoticed, for those where anger lies just below the surface, it may act as a trigger.

It is not just a matter of increased sensitivity to provocation, however. Impulsivity – the tendency to act without thinking – is also an important factor. When angry people find themselves in anger-promoting situations, they respond even more impulsively with anger.

To illustrate this, one study looked at the relationship between impulsivity and anger in a group of patients admitted to a forensic psychiatry unit for crimes associated with mental illness.

These patients were obviously at high risk of violent and/or non-violent behaviours.[9] Offences ranged from murder, assault and domestic violence, to theft, arson and drug-related crimes.

These patients were subjected to a psychological test called a 'Go/NoGo' task. In this test, the patients viewed a series of pictures, some with anger-related content, such as angry faces or people fighting, others with neutral content. Each picture had a blue or yellow frame. Depending on the colour of the frame, participants were asked to press a button or not press a button (the 'Go/NoGo' component of this experiment). For example, a blue frame around the picture should prompt to press a button, while a yellow frame means that the button should not be pressed. Those patients with higher levels of anger made more mistakes, pressing the button with NoGo images, implying generally higher levels of impulsivity: an impairment in reining in their instincts. Faced with anger, when the images surrounded by these coloured frames contained angry themes, their level of accuracy, their ability to suppress the impulse to press, was even poorer. This implies that in the angry, impulsivity is worsened by the emotional context. Anger makes you more impulsive.

When it comes to the underlying roots of anger within the brain, certain rare people provide valuable insights. One such person is 'DR', who had first suffered from epilepsy in her twenties.[10] Her seizures had proved difficult to control with medications alone, and brain surgery was proposed in an attempt to cure her of her convulsions. She underwent a series of operations to destroy her left and right amygdalae – almond-shaped structures deep in the temporal lobes of the brain – thought to be the anatomical sources of her seizures (Figure 1). The amygdala is a crucial part of a network called the limbic system, the circuitry within the brain most strongly associated with emotions (Figure 2).

After her surgery, DR exhibited some changes beyond simply the control of her seizures. The destruction of her amygdalae by the surgeon's knife had led to some unintended consequences. While she could recognise faces easily, she found that she had great difficulty interpreting facial expressions of emotion, particularly for fear, but also anger and disgust. While her hearing was normal, she had difficulties perceiving emotion from the intonation of speech. She could not hear anger or fear in the voices of speakers. Her ability to pick up on social cues expressing certain emotions, both through the facial expressions and voices of those around her, had been profoundly impaired.

Cases like DR's certainly implicate the amygdala in the perception of anger and fear, the ability to recognise these emotions in others. Other studies also link this brain structure in the generation of anger, not only in the recognition of it. In scans monitoring brain activity, this region lights up in response to angry stimuli, and different areas of the amygdala have been shown to play a role in expression of emotions, learning associated with emotions, and threat detection.[11] This explains why anger or aggression might sometimes arise in people with temporal lobe epilepsy, where seizures arise within or close to the amygdala, causing transient dysfunction of this important brain region.

When it comes to wrath, however, the amygdala is not the only player in the game. Many cases demonstrate that another area of the brain is intimately involved in the control of anger and aggression. One of the most famous of these cases is that of Phineas Gage, of whom I have written elsewhere.[12] Gage was a young man working on the construction of the railroads of the Eastern Seaboard of the United States. On a fateful day in September 1848, while tamping down explosives with a long, heavy iron pole in Vermont, he unfortunately ignited the explosive within a hole that had been drilled. The explosion drove the tamping iron through his skull, penetrating under his chin, passing through the frontal lobes of his brain, before flying out of the top of his head and landing some distance away.

Surviving his physical injuries, it was Gage's personality and habits that were most damaged subsequently. The change in him was rather dramatic. Previously a polite, considerate and God-fearing man, he became belligerent, angry and aggressive. Through Gage, and others like him, it became obvious that the frontal lobes have a role in regulating and inhibiting behaviour, including impulsivity, anger and other basic instincts. And, as we will see, in sexual behaviour too.

Introducing the pre-frontal cortex

There is a region of the brain in humans that is three times the size of that of our nearest living relatives, the great apes.[13] However, it is not necessarily the size of this brain region that distinguishes us from our closest animal cousin. Instead, it is how it is organised that contributes hugely to what it is that makes us human.

This lobe, the frontal lobe, extends from above the eyes to the midpoint of the brain (Figure 3). At its rear-most margins sit the motor areas, those regions directly controlling motor function, but of course our movements do not differentiate us much from other species. It is the area of the cerebral cortex immediately in front that is more notable – the pre-frontal cortex (PFC). The PFC comprises roughly a third of the entire cerebral cortex, the outer surface of our brains often referred to as grey matter. This is the area of the brain with most recent evolutionary origins, and also the last area to fully develop in our lifetimes; its development is only complete in late adolescence.

Why this brain area might be considered as the seat of humanity is down to the roles it plays. The PFC, through its wide connections to many other areas of the brain, is the navigator of the complexity of our lives. The PFC is crucial to decision-making, to reasoning, to the expression of our personalities, and to social cognition – those psychological processes necessary to participate and take advantage of living within a social group.[14] It is the crossroads of our internal and external experiences, the melting pot of our outside world and our inner selves. Our abilities in these neurological functions are what truly set us apart from the rest of the animal kingdom.

Within the PFC, there are further divisions, whose functions are intimately related to their connections.* One of these functional regions of the PFC is most relevant to anger, and indeed other behaviours. It resides in the deepest parts of the frontal lobe, its underside above the eye sockets, and deep in the middle of the brain. This area, called the ventromedial PFC (vmPFC), is more intimately connected to areas of the brain implicated in emotions, pleasure centres and drivers of basic instincts.

These anatomical connections reflect its functions. The vmPFC represents the nexus of those rational parts of the brain and those that fuel our emotions and our primitive desires. Damage to the vmPFC results in impairment of personal and social decision-making. People with injuries to this area of the brain will often have impaired moral judgement, have a weakened discernment of social norms and will often make decisions

---

* On the outside of the PFC – the region closest to the temple – is the lateral PFC (further divided into the dorsolateral (dlPFC) and ventrolateral PFC (vlPFC) – *dorso* referring to closest to the back and *ventro* being closest to the stomach). This lateral region is present only in the brains of primates, not other species. The dlPFC is the area most closely linked to what are termed 'executive functions', such as planning, abstract reasoning and working memory. It draws on inputs from areas of the brain central to movement, memory and sensation of all sorts. The vlPFC is thought to have an important role in attention, and cognitive reappraisal – changing how one thinks about a particular situation.

driven by immediate reward, just like Phineas Gage and other people we will meet. Damage to this region of the brain is most closely linked to a failure of inhibition of our behaviours, resulting not just in impulsive anger, but also excessive swearing, poor social interactions, hypersexuality, compulsive gambling or drug use, and an inability to empathise. It is the vmPFC that can most be considered our 'anger brake', and indeed the inhibitor of other antisocial behaviours.

While Phineas Gage and other historical cases of devastating brain injuries give us some insights, more modern studies are less reliant on tamping irons, explosives and the like. This more recent research, as with those cases of old, also show us the importance of the amygdala and the pre-frontal cortex in the experience of anger and the expression of aggression respectively.

One such study involved participants playing a game called the Inequality Game against other players, some of whom were cooperative, fair and pleasant, others being obnoxious and unfair.[15] During the game, participants were monitored for anger, but also for decisions in the game that might be seen to punish their opponent. When scanned in an MRI scanner assessing brain function rather than structure,* while viewing the faces of their fair and unfair opponents, the researchers found that the intensity of experienced anger correlated well with activity in the amygdala and other related regions of the temporal lobe. However, what they also found was that activity in two regions of the frontal lobes, including the pre-frontal cortex,† correlated well with the inhibition of punitive behaviour

---

* Functional imaging represents a variety of techniques to image the function rather than the structure of the brain. The most widely used of these techniques is functional magnetic resonance imaging, or fMRI. Utilising a powerful magnetic field, blood flow activity within the brain itself is measured as a proxy marker of neural activity; as regions of the brain become more metabolically active, regional blood flow responds to deliver more oxygen.

† The dorsolateral pre-frontal cortex and the anterior cingulate cortex.

in the game when playing 'unfair opponents'. This implies that these regions of the frontal lobes are important in the regulation of emotions, conflict resolution and inhibition of the consequences of anger. Indeed, other studies have also implicated further regions of the frontal lobes in the experience of anger intensity.[16] While anger, and its intensity, originates from the amygdala, it is the pre-frontal cortex that weighs up our response to it.

The hormone of aggression

From a psychological perspective, the study of anger is fascinating, but from a societal, medical and legal perspective, it is aggression – the behavioural consequences of anger – that holds the most relevance. It is aggression, not anger, that results in the estimated deaths of roughly 750,000 people in the world every year due to interpersonal violence (and that figure does not include those killed in armed conflict),[17] with countless other victims of non-fatal violence. Aggression is the foundation of many of the patients I have seen with catastrophic brain injuries, the consequences of punches, kicks or baseball bats to the head.

Aggression is also easier to study. There is a clear objective outcome rather than an internal experience of an emotion. Measurement does not depend on someone just reporting how angry they feel. It relies on overt acts.

Some people respond more easily with aggression than others. We only need to look around us, to walk down the street, read the papers or watch the news to bear witness to this. Why one person might respond to an anger-inducing incident with a stern word or even total passivity, and another with a fist or a knife, has long fascinated lawyers, psychiatrists, philosophers and teachers, given the impact of aggression in almost all spheres of private and public life.

One very easy answer to this question is that being male is the strongest factor, that testosterone is the explanation for aggression.[18] In almost all mammals, males are more aggressive than females (notable exceptions being lemurs and spotted hyenas). Testosterone enhances sensitivity to social threats. It activates centres of the brain involved in aggression and the processing of threat.

It may be that surges in testosterone associated with competition or threat may facilitate aggression to deal with those threats.* Indeed, testosterone levels have been shown to be directly affected by competition in males. Levels rise prior to competition, in anticipation of conflict, and swing wildly according to the outcome. If you win at wrestling, tennis or even chess (an activity not renowned for its sexual allure), your testosterone levels rise; if you lose, they fall. These findings have led some researchers to argue that the evolutionary function of testosterone is related to competition between males, with females being attracted to winners of these competitions. And if indeed this is the case, that mating success is the key outcome of the relationship between testosterone and aggression, then testosterone levels should fall when you have mated. This relationship between testosterone and mating has borne out: testosterone levels fall when a man marries, but rise again after a divorce, when he must compete again for female attention.[20]

This relationship between testosterone and aggression

---

* Testosterone is not the only hormone implicated in aggression. Cortisol, a hormone produced in stressful situations, also gives rise to increased aggression. Other hormones, like oxytocin, which has a role in enhancing social cohesion, have been linked to the behavioural coordination necessary for proactive aggression for the purposes of the group.[19]

appears to be important even before birth. Exposure of the growing foetal brain to testosterone is fundamental to the differentiation of the sexes. But even among individuals of the same sex, exposure to levels of testosterone within the womb varies.

A strong marker of the level of foetal exposure to testosterone is the ratio between the length of the index finger and ring finger, especially on the right hand, not only in humans, but in other species such as mice and baboons. Having a longer ring finger than index finger is associated with higher prenatal testosterone exposure, and lower levels of the female sex hormone oestradiol (oestrogen) exposure, and is linked to 'masculine' behaviour. The reason for this difference in hand development is uncertain but may be explained by genetic factors.[*]

This finger-length ratio is linked to a range of characteristics, such as verbal intelligence, numerical intelligence and being 'agreeable'. In men and women, the ratio is associated with likelihood of attacking in a simulated war game,[21] and levels of aggression after viewing an 'aggressive' music video.[22] It also associates with general levels of aggression, at least in males.[23] The longer your ring finger is relative to your index finger, the more aggressive you are likely to be.

These findings imply that exposure to high levels of testosterone prime the brain for more aggressive behaviour in later life, at least during the development of the foetus. The relationship between testosterone levels in adulthood and aggression is more complex, however. While fluctuations in testosterone levels may be relevant, absolute testosterone concentrations in adulthood do not correlate well with aggression, in males or in females.

---

[*]   Some genes have been identified that link to both gonad and hand development. Additionally, variants in genes that encode the receptor for testosterone have also been shown to influence this finger-length ratio.

Therefore, it appears that levels of testosterone in the chemical soup we inhabit in our mothers' wombs have a direct consequence on aggression, although the relationship between this hormone and wrath in adulthood is less clear-cut. And although testosterone is considered to be the 'male' hormone, females are exposed to it too.

While being male is one of the most important factors in the outward demonstration of wrath, this view of aggression being an almost exclusively male domain has been challenged in recent years. Males undoubtedly exhibit physical aggression more than females, but rates of verbal aggression between the sexes are similar, and females display indirect or social aggression – social manipulation intended to harm psychologically or socially – much more frequently.[24] This sort of indirect aggression can be seen in four-year-old children, but increases in frequency with age, since it requires a certain degree of social intelligence to analyse social situations and manipulate them.

The warrior gene

When it comes to reactive aggression – that impulsive response to threat or frustration, the type of violence that arises in the heat of the moment – it is not just down to testosterone. Genetics also clearly plays an important role. Twin studies – comparing genetically identical twins to non-identical twins, all of whom should have been nurtured in the same environment – suggest that our genes contribute about 40–70 per cent of our aggressive tendencies.[25] Identifying these genetic influences gives us pointers towards the possible chemical underpinnings of this aggression trait.

One of the first genes linked by researchers to aggression is called *MAOA*. This gene produces an enzyme that breaks

down chemicals in the brain. These chemicals are neurotransmitters that transmit signals from one nerve cell to another.
Specifically, this enzyme mops up neurotransmitters such as
serotonin, dopamine and, importantly, noradrenaline (norep-
inephrine), one of the main signallers of the flight-fright-fight
response, both within the brain and elsewhere in the body.
High potency of this enzyme results in a more rapid breakdown
of these neurotransmitters within the body and brain, and
hence lower levels of these chemicals. Low activity of the
enzyme allows levels of these chemicals to build up to higher
levels.

Severe mutations in this gene cause the MAOA enzyme to be
inactive, and thus dramatically boost the levels of these neurotransmitters. This results in a syndrome – Brunner syndrome
– which only affects males (due to the gene being located on
the X chromosome; women are likely to be carriers rather than
having the syndrome, since females will have one mutant version
of the gene and one normal version). Those males carrying
these mutations, and consequently with no or very low levels
of the enzyme, are highly prone to violent outbursts, triggered
by frustration, anger or fear, often leading to criminality (such
as rape, murder and arson).[26] As in humans, when this gene is
experimentally silenced in mice, they become extremely aggressive, with massive increases in certain neurotransmitter
chemicals – up to a ten-fold increase of serotonin, for example.

If these highly damaging mutations in the *MAOA* gene can
lead to such severe violence, then perhaps more common variants in the same gene may also influence human behaviour.
Indeed, it is apparent that these common variants, not just rare
mutations, also have an effect. Less harmful genetic changes in
this gene, modifying how effectively the gene produces the enzyme
rather than stopping the enzyme being produced altogether, are
also directly linked with many aspects of aggressive behaviour,

such as hostility, antisocial personality, increased risk of joining a gang, increased risk of using a weapon in a fight, and with increased impulsivity.[27]

These findings have led to the *MAOA* gene receiving the moniker 'the warrior gene' (or more offensively, 'the psycho gene'). There are even examples of defendants for murder or attempted murder having this gene sequenced, in an effort to have their sentences reduced, with some success.[28] In one case from the US, a conviction was reduced from first-degree murder to voluntary manslaughter, and in another case from Italy, a sentence was reduced from life to twenty years, on the basis of the defendants' *MAOA* genetic sequences. Essentially, the defence argument is that the genetic make-up of the accused leads to a diminishment of responsibility for their actions.

MAOA is not the only gene implicated in the genetic origins of aggression, although it is the most studied. Variation in genes involved in serotonin regulation – transport of this chemical in and out of cells, or detection or production of it – have also been implicated in aggression, but also more widely in emotional regulation. Other genes that regulate dopamine metabolism or dopamine receptor function have also been associated with impulsivity, violent conduct and even cases of murder.[29]

Violence begets violence

Thus, our levels of aggression are, at least to some extent, a part of our genetic inheritance. Our propensity to violence is to some degree gifted to us by our parents, like our hair colour or the shape of our nose. But these genetic factors, especially the ones that we know about, are only a fraction of the picture.

As with all areas of biology, there are questions of nature versus nurture. Our parents can also contribute in other ways, not just through our genes.

In patients, one group of individuals prone to anger management issues are those with borderline personality disorder (BPD), also known as emotionally unstable personality disorder. The essence of personality disorders are extremes of normal personality traits, magnified to the extent that they cause harm or distress to the individual or those around them. BPD is characterised by huge and rapid fluctuations from the positive to the negative, when it comes to self-image and relationships with others. In practice, this often manifests with anxiety, irritability and mood disturbance, as well as very impulsive behaviour, influencing spending, the misuse of substances or sexual activity.[30]

BPD has costly consequences. This disorder is associated with the inability to hold a job, high rates of other mental health conditions and extremely high rates of suicide (up to 6 per cent in one study). Some 75 per cent of people with BPD will report suicide attempts. People with BPD have a significantly shorter life expectancy. And BPD is not rare; some 1–3 per cent of the adult population have it.

An additional feature of BPD is inappropriate intense anger, difficulty reining in aggression, with frequent displays of temper or getting into frequent physical fights. This propensity for impulsive violence is not insubstantial. One study found that 73 per cent of people meeting diagnostic criteria for BPD had been involved in violent incidents over the preceding year.[31] Compared to those without BPD but with other psychiatric diagnoses, those with BPD were 62 per cent more likely to commit seriously violent acts, and more than twice as likely to commit aggressive acts. The researchers reported hitting with a fist or object, shoving or pushing at one end of the spectrum, and assault with a deadly weapon or rape at the other extreme. However, they pointed out that there is significant overlap between BPD and other antisocial traits.* A more frequent problem in clinical practice is aggression directed towards oneself, turned inwards, contributing to those high rates of self-harm and suicide attempts.

The origins of personality disorders are not fully understood. However, a major risk factor for developing them is 'childhood adversity'. This term is often used as a euphemism for childhood trauma – physical, sexual or psychological abuse – but in the broader context of personality disorders may include bullying or violence at schools, for example.[32] When it comes to BPD, however, this type of personality disorder more than any other, is associated with abuse and neglect, found in between 30 and 90 per cent of people with BPD.[33] We see this day in and day out in our clinics. People whose childhoods have left an indelible mark on their brains, unleashing patterns of behaviour that blight their adult lives and those of the people around them.

Indeed, a growing body of work suggests that childhood adversity fundamentally influences brain development, on a structural, functional and neurochemical level. Our upbringings may profoundly affect our adult brains. How this might happen remains uncertain, but a number of different lines of investigation are currently being explored.

Firstly, these sorts of childhood experiences expose us to intense acute and chronic stress. Our usual response to stress is mediated by a major system in the body and brain, the

---

* Many of those individuals carrying out violence or aggressive acts also had elements of antisocial personality disorder (ASPD), which has significant overlap with BPD, with impulsivity, irritability and anger, and so the nature of the relationship between BPD, ASPD and aggression remains blurred.

hypothalamic–pituitary–adrenal (HPA) axis. Acute stress causes the hypothalamus to release hormones that ultimately trigger the adrenal glands, small nubs of hormone-producing tissue above our kidneys, to produce cortisol. This hormone acts throughout the brain and body to prepare for a stressful event. Cortisol release changes our metabolism, our immune function, our thinking and behaviour, all with a view to the expectation of injury, harm or running away. It serves a very useful purpose in that moment of acute stress or threat. If levels of cortisol are heightened in the long term, however, in the context, say, of the stressful environment of an abusive home or school, this may ultimately result in changes to brain structures involved in emotional regulation.

Childhood trauma may also change levels of neurotransmitters or the formation of connections between nerve cells within the brain – synapses. Many of these chemicals we have already encountered – noradrenaline, dopamine and serotonin, among others. Other chemical changes encountered in individuals with BPD include growth factors that control the development of synapses. For example, the brains of people with BPD have altered levels of brain-derived neurotrophic factor, a crucial molecule that mediates brain development, and is associated with stress.

There is an additional explanation for how our childhood environment may fundamentally affect our development. Essentially, your environment can influence how your genes function without intrinsically altering your gene structure. Instead, your environment can modify how your genes behave, by increasing or decreasing, even switching on or switching off, gene activity. Environmental factors may influence how our genetic code is translated into our physiology and anatomy. The architect's blueprints, our genetic sequences that are the essence of us, are drafted in indelible ink, but once on the construction site, scribbles of the builder's notes gradually cover the pages. This phenomenon of our environment influencing our genes is termed epigenetics.*

In rodents, primates and humans, there is increasing evidence that stress in early life induces chemical changes in genes that affect behaviour and brain development. In rodents, for example, the quality of maternal care is associated with changes in a gene important in the development of the brain.† These alterations occur in regions of the gene that regulate how active that gene is – so-called promoter regions. The nature of maternal care given by rat mothers therefore directly affects the genetic machinery in their offspring.

These sorts of findings have been replicated in humans, with similar changes in the same gene in those with a history of early life adversity.[34] Analysis of the genetic code of individuals with BPD demonstrates epigenetic changes in several genes implicated in brain development when compared to control subjects, suggesting that these changes may contribute to the development of BPD in the context of childhood trauma.[35]

Therefore, adversity appears to alter the chemistry and development of the childhood brain. But how do they influence the adult brain's function? These stress-mediated changes in brain

---

* Our environmental factors can influence our genetics in several ways. One major epigenetic mechanism is that of DNA methylation. Specific sequences of genetic code can acquire molecules of carbon and hydrogen called methyl groups, altering the molecular structure of the bases that constitute the DNA structure. The result of this methylation is to modulate the reading and expression of a particular gene. This process of methylation and demethylation is thought to be strongly influenced by environmental factors. Indeed, methylation is thought to play an important role in the aging of our cells.

There are other major mechanisms of epigenetics. DNA is wrapped around proteins called histones, and the structure of these histone proteins can be modified by environmental factors. Such modification may influence the accessibility of genes to the machinery involved in their expression. Other sequences of genetic material, termed non-coding RNAs and messenger RNAs, may also be altered by environmental factors, and these molecules also have effects on the process of producing proteins from the DNA sequence.

† This gene is called *NR3C1*.

development may explain why patients with BPD show subtle differences on brain imaging too. Unsurprisingly, these changes are mainly located in the limbic circuitry and the pre-frontal cortex, those regions most associated with emotional reactions and impulsivity respectively. The most consistent changes are within the amygdala – the fear centre – and associated structures, which are generally smaller in people with BPD.

Neuroscience is therefore gradually exposing the links between childhood neglect and abuse, our genes and our brain structure and function in those individuals with BPD. Even beyond those with a formal diagnosis of a personality disorder, however, aggressive people also exhibit differences in these brain areas. Several studies using a variety of imaging techniques show differences in these familiar regions of the brain involved either in the regulation of emotions, or in the inhibition of behaviours – the limbic system and the frontal lobes.[36] And as with BPD, even for those without a personality disorder, environmental factors predisposing to aggression include exposure to community aggression, neglect, abuse, and domestic violence.

This makes perfect sense. If you grow up in a violent world, to be highly attuned to signs of violence, to be vigilant for it, to pre-empt it, and to meet aggression with aggression, seems eminently prudent for your survival. To augment those aspects of your nervous system to adapt to your environment is a natural response. When it comes to wrath, nurture may be as important as nature.

The interplay between our genes and environment

All this begs the question: why do some people exposed to these risk factors, either genetic or environmental, become aggressive or develop a personality disorder, while others end up leading very normal lives? One possible explanation is the interaction between these two factors. The effects of our environment may depend upon our genetic background, and vice versa.

Our genes may protect us from the effects of our upbringing, or make us more vulnerable. For example, boys who have suffered abuse but also have a high-activity version of the *MAOA* 'warrior gene' (and thus have lower levels of the neurotransmitters that mediate anger, due to their more rapid breakdown) are less likely to develop antisocial tendencies in adulthood. Those with a low-activity version are much more likely to end up with convictions for violence.[37]

Variants in other genes influencing serotonin and dopamine transmission have also been found to interact with environmental factors to influence likelihood of aggression, sometimes in rather complex ways. One striking study showed that variants in another gene implicated in violence (a gene that influences the regulation of serotonin) were associated with carrying a gun in pre-9/11 USA, whereas after the terrorist attacks, gun-carrying levels were similar irrespective of what genetic variant people possessed.[38] This implies that, while the presence of environmental factors like stress or violence may give rise to aggression in genetically predisposed individuals, the nature of those environmental factors, their magnitude or severity, may also moderate the effects of our genes on our behaviour. People with this gene variant, who were more likely to carry a gun prior to 9/11, have a greater likelihood to respond to everyday situations with a need to protect themselves. However, when the general perception of threat is elevated beyond the everyday, those without the variant behave no differently.

* * *

It is ironic that aggression results from early exposure to aggression, but also that, on a neurobiological basis, it is due to a form of neurological conflict. The perpetual battle between those areas of the brain that drive and regulate our emotions and our deeds. In the context of aggression, one circuit within the brain, the limbic system, is involved in the generation of anger, the ascribing of stimuli of frustration or threat with that emotion. The other is the frontal lobe – most specifically the pre-frontal cortex – which acts as a brake, an inhibitor of the behaviours associated with emotions, and acting upon them. Under normal circumstances, these two systems are in perfect balance, allowing us to experience the motivating force of anger without resorting to aggression or violence. When out of alignment, however, with an underactive pre-frontal cortex or an overactive limbic system, and a failure of inhibition or an exaggeration of emotional response respectively, aggression and violence rear their ugly heads.[39]

The origins of this imbalance between these regions of the brain may be genetic or environmental. Sometimes though, the cause is more transparent, more evident, more immediate. Like Phineas Gage and his tamping iron flying through his frontal lobe, or many others with overt neurological dysfunction.

For Tom, it was not a devastating head injury that changed him. Instead, it was exercise that transformed his life, in an unexpected way. 'I was warming up for a run,' he tells me. 'Quite early in the morning. I was skipping on the patio. My eye just started to feel a bit weird, and then I got this really awful pain that just stopped me. A pain round my head. I felt like something was really wrong. I was like: "Fuck, that is really, really bad." I've got a reasonably high threshold for pain,' he says, 'but I remember feeling like shit. "Whatever that is, it is not just that I haven't drunk enough water today." And then I started throwing up, and kept throwing up with this headache. I just had this unbelievable pain in my head. My mum was there – she is a worrier at the best of times – and she knew there was something really wrong with me. She called for an ambulance.'[40]

On this day in April 2017, Tom was in his late twenties, working in music. From his early teens, he had always wanted to be in a band. He had started playing the drums and guitar. 'Father told me I wasn't allowed to join a band until I was sixteen. So [on my sixteenth birthday], I joined a band with older people. In sixth form, I was playing all over the UK on the weekends, sneaking into venues while still underage.' Initially it was rock music, then punk. 'Lots of squats and very interesting people. Crusties. People who looked like they hadn't had a bath for quite a while. A lot that was quite colourful. But I just fell in love with it. Despite looking crazy, and being into this really aggressive music, everyone was really kind, and really supportive. You would end up being part of this broad community.'

Tom had enrolled in a degree in illustration, although he describes it as a ruse, as he spent most of his first year touring with his band. He left in his second year, to pursue his dreams of musical success, and began to utilise his illustrating skills by designing record sleeves and merchandising for other bands as a 'side hustle'. After several years of an itinerant life with the band, however, Tom sought a bit more stability. At the age of twenty-three, he took a job in a music studio. He started off as an assistant, fetching tea, cleaning the studio. 'I was sleeping under the piano. It all sounds very "rock and roll",' he scoffs, signing the inverted commas with his hands. 'I look back with rose-tinted glasses now, but it was very brutal at the time. It was a really hard way to live. But I think that's part of what I found really enjoyable: that sense of adversity.'

He rapidly worked his way up, and soon began to play as a session musician and sound engineer for bands recording their albums. Tom would try his hand at anything and would also photograph bands and concerts. As he reminisces about these years, I can hear the pride in his voice, but it is tinged by some self-analysis. 'I thought to myself: "This is actually living. I'm doing a real thing here." To be honest, I was probably a bit of an asshole.' He describes a righteousness, a feeling of superiority, that he was living a 'true life', while those around him, with nine-to-five jobs, were just on the treadmill. 'It took me a long time to get out of that mindset. It was just so destructive, and intolerant of others.'

His 'true life' came to an abrupt halt with that call for an ambulance, and the aftermath. Tom recalls dipping in and out of awareness on the ride to the local hospital, still vomiting and in incredible pain. In the emergency department, a scan revealed bleeding. Not within the substance of the brain, but into the cerebrospinal fluid that bathes the brain – a sub-arachnoid haemorrhage. Typically, this is caused by rupture of an arterial aneurysm, a ballooning of a blood vessel that causes the arterial wall to weaken and burst under the high arterial pressure. The resultant sub-arachnoid haemorrhage is a medical emergency and is potentially life-threatening. The sudden increase in the pressure inside the skull can cause damage to the brain itself, can jeopardise the brain's ability to regulate breathing and heart rate, and the blood bathing the arteries supplying the brain substance can precipitate spasm of the blood vessels, resulting in secondary strokes. 'There was a big fuss, because they had accidentally given me ibuprofen, which can make you bleed.' Tom was rapidly transferred to the nearest university hospital, with neurosurgeons on standby. 'I was almost relieved that someone else was taking the reins, even if I was dying. I was quite cool with it all. I remember a doctor leaning over me in the bed and saying, quite dryly, "You could die."' Repeated scans of the arteries in his head failed to find the source of the bleeding, however, something seen in a minority of cases of sub-arachnoid haemorrhage. With nothing evident to fix, the neurosurgeons were stood down, and Tom was treated with medication.

Tom lay in his bed in a neurosurgery ward, terrified by others around him in a more parlous state. People with major brain trauma or brain tumours. Patients whose lives teetered on the edge, or whose devastating injuries were causing confusion or aggression. He recalls being desperate to go home. Then, four-teen days later, he was told that he could leave if he wanted to. 'I thought: "Fuck this, I'm just going to go." But I had no concept of how [unwell] I actually was. They just waved me out of the door. It was: "Bye-bye."' In hindsight, Tom regrets his rapid discharge. He feels he missed an opportunity to learn more, to recover more. A follow-up appointment several weeks later, he summarises as, '"Oh, you seem fine." That was it.'

Tom was not fine. He was walking and talking, overtly fully functional. But it was a very different Tom. 'In a matter of weeks, I had changed from being very aware of myself, being in command of myself, to realising that I could not organise my thoughts. Looking in a mirror for the first time was really unusual, because I felt my appearance had really changed.' He describes a profound sensitivity to light and sound, an inability to tolerate the morning sun or the day-to-day noises of life.

But he also reports another type of sensitivity. As he speaks, he pauses briefly. 'Not to be too "woo-woo" about it,' he chuckles – Tom uses humour throughout our discussions, making light of his situation as he tries to make sense of it. 'It sounds daft, and I am still quite guarded about it. But I also felt really sensitive to people's emotions. If someone was speaking to me, with a sense of urgency, not even aggression, it would immediately trigger a fear response. I would feel over-whelmed.' I ask him to give me a specific example. 'Obviously an argument would be pretty bad. But even if my mother was making me a coffee. I would feel like I could sense her thinking [whether to use dairy or oat milk]. She has always been an overthinker, a worrier. And because it was about me, I felt really overwhelmed. I really needed things to be simple, simplified, for her to remain as neutral as possible. So, when she would do a little thing like that, despite the fact that she has always [been like that], it felt like she was projecting her worry onto me. But really, she just got the milk wrong.'

These frustrations, imperceptible to others, would trigger something in Tom that he still does not understand. 'I would feel a flash of anger. It was somehow inconsiderate for her not to be in better control of her own neuroses around me. Which doesn't stand any scrutiny to me. I don't think that it is fair for me to do that.' That flash of anger, particularly in the early days, was also accompanied by fear. A feeling that Tom did not know what was going to happen. A feeling of being at the absolute limit of his capacity to cope.

Over the first few months, Tom felt like he was making good progress, although now he ascribes this to burying his head in the sand. 'I tried to return to normal life without acknowledging any difference in me. I tried to steamroll my way through it all. I came crashing down to earth a few months later. I found a lot of my relationships were breaking down. I was just a dick.'

Despite the cause of Tom's behaviour being likely structural damage to his brain rather than a drug, there are very clear parallels between what he and what Jono, with his Keppra rage, experience. Tom describes an anger that wells up from deep inside him, from calm to a red mist descending within a split second. Pre-haemorrhage, Tom describes himself as 'a bit cowardly'. 'I was actually a bit of a wuss really. I was never into fighting or anything like that. That's the thing about aggressive music. A lot of people [into that style of music] are not aggressive. Perhaps that's how they get it out.'

After the haemorrhage though, it seems that his innate 'coward' has been overcome by sheer rage. One of the first episodes occurred after he had taken a job to get back into the workplace – at a T-shirt printing business that Tom says was a fantastic place to work: 'full of freaky alternative people who were just wild. It was really cool. Just a fab opportunity for someone like me.' His cognitive faculties after his haemorrhage were clearly not fully recovered, and he was aware of making small mistakes. In his previous life, he had been highly competent, and took pride in it, so he found this a little humiliating. His humiliation increased several-fold when he found out his boss was making denigratory remarks about him to others. 'I accosted him in the office. I said: "If you ever say anything like that again, I'm going to batter you." The blood was just pumping. I remember looking round the office for something to hit him with. He was much bigger than me, and logically I know this is not a great way to deal with something like that. But could I control myself in that moment? Absolutely not.' Tom says he felt terrible about the incident afterwards. 'I felt like a bully. I felt really bad about it. To be fair to him, he took it pretty well,' he smiles ruefully.

His anger has also put him at physical risk. He needed access to the front of his house to get his drums into the car, when he found it blocked by six builders working nearby. 'I just completely lost it. I was just screaming at them, threatening them. I remember shouting: "You think you fucking own the place so you can do whatever you like! You think I don't matter! You don't care about anything else! I'm going to slash the tyres on your van." It was totally ridiculous.' I ask him how they responded. Tom admits that they could easily have beaten him up to within an inch of his life, but luckily for him, they just looked at him dryly, and ignored him. As the anger subsided, Tom was overcome by embarrassment and shame. 'I just thought: "You fucking idiot. They're outside your house now for a week." I just hid as much as possible. I just felt really wrong for [behaving in such an aggressive way].'

A complex relationship with brain injury

As Tom had lain recovering in the neurosurgical unit, surrounded by other patients with profound and frightening behavioural changes due to brain injuries, he had shared more than just a ward with them. While many of those individuals occupying the beds adjacent to him had suffered brain damage due to trauma – the impact of head on concrete, on car bumpers, or on fists – Tom had sustained a similar sort of damage. In Tom's case, however, the insult had come from within rather than without, from the rupture of a blood vessel, and the associated harm of arterial blood surrounding and irritating his brain. The consequences, regardless of the mechanism of injury, were somewhat alike. A wholesale disruption of the delicate dance between those areas of the brain that induce anger and those regions that tame it.

And while wrath can arise from both genetic and environmental factors, in the world of neurology, one of the commonest causes is brain injury – in particular as a result of trauma. Many patients I see in my clinics have been victims of traumatic brain injury (TBI), sometimes a minor slip or fall down some stairs with concussion; others with massive injuries due to road accidents, weeks on intensive care, multiple fractures and major neurosurgery. It is very common outside my clinic too, for TBI is the single largest cause of death and sickness in children and young people.

Until the age of five, boys and girls are equally at risk, but in adolescence and young adulthood, it is males who put themselves much more in harm's way, with a doubled risk compared to females. Due to the anatomy and mechanics of the brain (a blow to the head puts shear stresses on particular areas of the brain, and bony protrusions around the base of the skull make some areas more liable to bruising or lacerations), the frontal regions are particularly vulnerable to damage. Even mild injuries can cause particular changes in thinking – problems of attention, impairment of inhibition, impulsiveness and poor social judgement.

These alterations in mental functioning are potentially a highly toxic combination. Mix impetuousness, an inability to suppress inappropriate behaviour and a diminished capacity to weigh up social rules, and it is easy to see why this might predispose to aggression. Indeed, these sorts of cognitive impairment associated with TBI have perhaps unsurprisingly been linked with aggression, violence and crime.[41] TBI, especially in childhood, may disrupt normal brain development, particularly for those systems that influence social interaction, and lead people into more aggressive modes of operation. Studies following up children with TBI have found that, even in mild injuries, personality change is common, with rapid and exaggerated changes in mood, disinhibition (the inability to withhold or suppress unwanted or inappropriate behaviour) and resultant aggression, and that these changes are associated with damage in the pre-frontal cortex.[42] Head injury in childhood has been associated with an increased risk of dropping out of school and violence, even when other factors are taken into account (since an alternative explanation is that kids who are impulsive and risk-takers in the first place may actually be more at risk of head injury).[43] These kinds of findings are seen in individuals acquiring brain injury in adulthood too. Vietnam War veterans with injuries to the frontal lobes have been found to be more aggressive and violent compared to those without injuries, or injuries elsewhere in the brain.[44] Selective damage to the frontal lobes, in childhood or adults, sets the scene for violence, by hampering an ability to suppress it.

When either general populations or offenders are examined, a link between TBI and crime generally stands. Studies from several countries, including England, New Zealand and Finland, show that TBI is linked to mental health disorders, substance abuse and criminality, especially violent crimes.[45] One very large study, using Swedish national patient registers over a thirty-five-year period, showed that 2.3 per cent of the population had committed violent crimes. This rate was over three times higher for individuals who had suffered a TBI. Even when these TBI cases were compared to siblings who had not experienced TBI, but presumably had similar genetic, economic and family backgrounds, TBI was associated with twice as high a likelihood of violent crime.[46]

In incarcerated young and adult offenders, rates of TBI are much higher than compared to the general population, and brain scans in violent prisoners are significantly more likely to show abnormalities, compared to non-violent prisoners.[47]

At present, the links between TBI, criminality and violence are yet to be fully elucidated. TBI may simply be a marker for people who are inherently more impetuous, risky or indeed predisposed to violence. A violent offender is going to be at higher risk from a brain injury due to their tendency to violence. However, some of the data relating to young children would point away from this explanation, and the personality changes seen in some people with TBI would support the view that TBI may give rise to criminal behaviour.

This area of research is important. A 2016 analysis of the economic costs of mild or moderate TBI in young people put the figure at about £155,000 ($195,000) per case, of which £60,000 ($75,000) related to costs associated with offending. For someone already in the legal system, these predicted costs rise to £345,000 ($440,000).[48] If the effects of TBI on subsequent criminal behaviour can be moderated, through rehabilitation or other treatments, or the initial injury can be prevented, the personal and economic cost implications may be huge.

However, there is one additional factor that acts as fuel to this fire. It is readily observable on a typical Friday or Saturday night in the city centres of the UK – alcohol.

The relationship between alcohol, traumatic brain injury and aggression is far from straightforward. Excessive alcohol use is associated with a tendency to impulsivity and personality disorders. Those who consume it to excess may in any case be at increased likelihood of aggression or putting themselves in harm's way. The impairment of judgement that is a hallmark of acute intoxication puts you at risk of falls, of road-traffic accidents; of situations that predispose to TBI.

Whatever the precise nature of this relationship, the role of alcohol in violence cannot be understated: in 2011, 73 per cent of homicides in the US and 53 per cent in Russia were alcohol-related.[49] Its links to domestic abuse and non-fatal violence are equally strong. Alcohol exerts a particularly toxic influence on the brain when it comes to violence – the loss of emotional control alongside its disinhibiting consequences. Alcohol is the catalyst of confrontation and erosion of self-regulation. Regardless of the links between excessive alcohol use and personality traits that might predispose to wrath, it also directly disrupts the function of the pre-frontal cortex.

Current data are inconclusive as to whether alcohol in and of itself is sufficient to cause violence, or simply increases the risk of it in susceptible individuals.[50] However, in addition to the direct impact of intoxication on the frontal lobes, chronic alcohol dependency is associated with chemical changes in the amygdala, causing this region to be more excitable. This suggests

that chronic alcohol abuse may directly induce brain changes
that make people more prone to aggression, independent of
acute intoxication.

* * *

As Tom continues to describe the life-changing consequences
of his haemorrhage, the alterations to his emotional state and
personality, Han, his girlfriend, sits quietly next to him. She has
never known the old Tom. Their relationship started a few
months after his bleed. They matched on a dating app as she
was on a train passing close to where Tom was living at the
time. They have been living together for the last three years.

As with Jono and his Keppra rage, Han describes parking
and driving as being regular flashpoints for Tom's rage. They
live in a picturesque cottage surrounded by rolling fields. 'Around
here everyone is a pensioner. Car parks are a clear trigger. Their
minds are not focused on others. But how could you verbally
abuse a pensioner? It would be him shouting, "Fuck off", or
"You old bastard!"' His rage has also impacted their relationship. 'I don't think Tom is going to be [physically] aggressive
towards me or anything like that. But it has made me not want
to say things. I think the issue would arise if I were overtly
emotional or angry. I have to spend some time with what I'm
thinking, so I can get it out in a much more calm and unemotional way. Keeping it quite logical, still based around feelings,
but without too much emotion attached to it.'

Han also describes some important differences, however, when
compared to Jono's case. In addition to the exaggerated fear
response that Tom experiences, intense shame features strongly.
Profound mortification at his outbursts, deep disgrace at his
inability to cope with what life throws at him. I am struck by
the fact that, despite the intensity of Tom's rage, it has never

escalated to the point of actual physical violence to Han or others, that his anger has not managed to override his intrinsic personality. Objects have incurred his wrath: walls, or mobile phones – 'I have smashed five or six now' – and, though there have been some close calls, people have never incurred physical violence.

There has been self-directed violence, however – self-loathing is his response to his changed character, intrinsically tied to his feelings of shame. In his crusade for 'normality', he has trained to be a Samaritan.* His very first call was someone trying to die by suicide. 'I was okay with it. But I was so ashamed that I hadn't told them I was unstable. I got myself really upset and angry. I had something glass in my hand, and I smashed it into my head, and cut my head open.'

On another occasion, Tom had signed up for an international music tour, only to realise at the last moment that he was incapable of managing it. His anger and his humiliation about his inability to cope expressed itself in him suddenly grabbing a knife and cutting his wrist. Fourteen stitches were required. He told his fellow band members that he had been mugged, and they posted this on their social media networks. 'I became embroiled in this lie. It was just awful.'

At the heart of these behaviours that we term sins is the concept that by our actions we are judged. Yet it feels like Tom is his own harshest judge. Despite attributing the new Tom to the damage his brain has sustained, it is obvious that he still blames himself, that he sees his behaviour as a moral failing on his part. 'I know I am capable of thinking logically and critically about things,' he says. 'The fact that I can't get a solid grip on this, I feel like I am letting myself down. I feel like I could be doing better. And that causes me shame.'

---

* The Samaritans are a UK and Ireland-based voluntary organisation, whose mission is to provide support for people at risk of suicide or in emotional distress. Volunteers staff telephone helplines.

Over the last few months, Tom has become more accepting of his situation. 'But I can still fall back into shame. The sense of loss is just enormous. It feels like proper grief.' I ask him if he feels the old Tom is lost, if that is what he is grieving for. 'Yes, I think so. I used to have a vision of my future. I could follow my internal voice. It was fearless. But that voice has now been silenced.' His need to control his environment, to avoid any triggers for his emotional sensitivity, has narrowed his world. 'I'm quite reclusive now. Between Han and my family, you are probably the only other person I have spoken to for months. I find it very difficult to engage with old friends. On the surface, I look the same, I look like me. But out in the world, I can't control everything. It is better for me to be away from it really. It is quite sad, but it is better for me to do it this way.'

While both Tom and Jono, with his Keppra rage, have some similarities, the differences in terms of their impact on their lives is evident. Despite Jono's outbursts, his life goes on much as before. He continues to work, have a social life and engages fully with the world around him. For Tom it is very much otherwise. He has withdrawn from the world in an effort to mitigate his behaviour; a response to the shame and embarrassment that he feels at his inability to control himself. Tom strikes me as more sensitive, more fragile, in contrast to Jono's robustness, though it is difficult to know if this was the case before his brain bleed, or as a result of it. It may also be that a drug has more specific effects, and that the bleed has not only had a direct effect on Tom's anger, but also his mood, anxiety and other aspects of his personality.

Evidently a profound transformation has occurred in Tom's brain with that sub-arachnoid haemorrhage. He has met a neuropsychiatrist, but this was early on after his bleed, and the main focus was his cognition – issues with his memory and mental fogginess. I am shocked to hear that he has not had any follow-up brain imaging, but not as surprised as I should be. Our healthcare systems are much better at diagnosing and treating more tangible diseases; less well set up and less funded to deal with aspects of illness that are less clearly defined. While some of his symptoms could undoubtedly be ascribed to the emotional trauma of the brain bleed, to facing one's own mortality at a tender age, to me his story is very suggestive of brain damage. Of cognitive dysfunction, light and sound sensitivity, all starting acutely after his bleed. His exaggerated wrath, and his inability to contain it, is not in isolation. His fear response is rendered susceptible to the slightest trigger: a reactive aggression unleashed both by an easily triggered underlying cause – of threat, of peril – and an incapacity to suppress its response. Without further investigations, however, understanding the true nature of his change remains uncertain.

Tom has resigned himself to never regaining full control again. Instead, in the last year or two, he has been seeking out a better understanding of the world that allows him to fit into it as he is. He speaks of a desire to find some inner peace, rather than trying to fix himself or control the world around him to prevent these outbursts. 'I have found ease in spirituality,' he says, 'helping me to feel a little happier, to enjoy day-to-day life a bit more. Music has been my religion; I have been praying to it my whole life. But now I can't expect the same thing from it. I seek that togetherness you get from a group of people working to produce something. I recognise that in organised religion – people sharing the same views, being sympathetic to the world around them, and me potentially feeling part of something. Something that doesn't fill me with shame about what I used to be or what I am now.' He hesitates, then laughs: 'But I don't know if that makes me sound like a wacko!'

An icy wrath, with an evolutionary imperative

The aggression that Tom and Jono show is of a particular type, a reactive aggression triggered in response to an immediate threat or stressor. It is impulsive and defensive, and is characterised by a strong emotional component. It serves no reward other than the here and now. There is no other benefit beyond the defence of self or the group, or of resources. Examples of reactive aggression are violence in the face of intruders, triggers such as insults, violent body language or territorial aggression.

But there is one aspect of wrath that we have only barely touched upon. For when it comes to the intent to cause harm, humans are capable of both reactive aggression and premeditated, cold-blooded, proactive aggression. While many acts of violence may have elements of both types, they are essentially distinct.

In stark contrast to reactive aggression, proactive aggression is predatory, offensive, without the 'heat-of-the-moment' aspect. It unfolds with calculated intent. Preparation for this sort of violence may lack anger or other emotions, and has a cognitive component to it: plans, tactics, calculating when the victim is vulnerable, maximising the power differential between the aggressor and the victim. It manifests as wars, bullying, sexual coercion, domestic abuse, premeditated murder or infanticide, and predation. And crucially, there is benefit beyond simply the removal of threat: the reward of status enhancement, power, wealth or sexual opportunities. A steely, frigid violence, with a greater purpose, far beyond the limits of mere survival.

By and large, we humans are a peaceful species, often able to cohabit in large communities in a non-violent manner. Yet, simultaneously, we possess a frightening propensity for violence, murder and mayhem. This apparent contradiction has been much debated. Some argue that the human race is intrinsically pacifist, but that violence is a function of our culture, while others propose the diametrical opposite: that our violent nature is reined in by cultural restraints. However, it appears that this duality is a little more complicated.

While our ability to murder on a massive scale, to organise and work with others to inflict death and destruction, is not in question, some researchers suggest that our intrinsic predisposition to reactive violence or aggression is extremely low indeed. Observations of pre-industrialised societies have found little evidence of reactive aggression at all. For example, ethnographers have observed the Aché people of Paraguay for decades, without seeing even very minor acts of reactive violence.[51]

When compared to our nearest relatives, chimpanzees and bonobos, we have a very different violence profile. Chimpanzees of both sexes show a high tendency to reactive violence. They also exhibit proactive aggression towards others and prey at levels similar to those seen in humans, sometimes in a coalition towards other chimpanzee groups and occasionally resulting in death. In contrast, bonobos are slightly less reactively violent, at least when it comes to male violence towards female bonobos. Bonobo family groups engaging in lethal warfare is unheard of.

Chimpanzees therefore have a high predisposition to both reactive and proactive aggression; bonobos have an intermediate level of reactive aggression and low levels of proactive aggression, whereas we humans have low levels of reactive aggression but very high levels of proactive aggression. These different violence profiles in such closely related species begs the question as to how these traits evolved. What are the evolutionary pressures, the difference in circumstances, that led us and our primate cousins to these different behavioural destinations?

One intriguing theory seeks to explain these differences.[52] In non-human primates, there is an obvious advantage to being aggressive in a reactive setting. High levels of reactive aggression favour individuals, especially males, in status contests and mating success, thus encouraging the spreading of genes that then further promote reactive aggression. So why should this tendency to reactive aggression have been bred out in us, unlike our chimpanzee cousins? What are the evolutionary pressures in humans that selected against these genes?

The answer may come from people like Caligula, Nicolae Ceaușescu, Muammar Gaddafi and many other feared and hated alpha males throughout history. Ultimately, these alpha males are vulnerable to less powerful males, through the human cognitive ability to form an alliance, to coalesce as a coordinated group, and kill. At first glance, this sounds counterintuitive: our human ability to kill each other lessens aggression.

According to this theory, capital punishment, execution through collaboration, has been a widespread practice throughout human history as a tool to control reactively aggressive males, and to a lesser extent, proactive aggression that is particularly destabilising to human society (such as infanticide, sexual coercion and grabbing other people's resources). Its ubiquity throughout all cultures suggests that it has been a consistent feature of human society, at the very least since humans spread out of Africa 60,000 years ago. Thus, capital punishment, the killing of those males whose levels of aggression are elevated to the point where they hinder societal cooperation and stability, proved to be the evolutionary driver for a lessening of reactive aggression in males.* While chimpanzees possess the ability to conspire and coordinate violence against neighbouring groups, they are incapable of killing the alpha male within their own group. A critical difference may be the need to use language to coordinate execution. While chimpanzees vocalise to coordinate

---

* Thankfully, modern society now has alternatives to capital punishment, and can deal with these alpha males in a different way.

hunting or fighting against other groups, they do not possess the communication skills necessary for the sophisticated coordination required for organised murder.

However, even in a species that has strikingly high levels of proactive aggression, there are some individuals that sit at the extreme. There are people who walk among us for whom proactive aggression is a modus operandi, where violence surfaces easily and without emotion. In the popular imagination, psychopaths are mentally ill, raging, slathering lunatics. But psychopathy is not a mental illness, in the sense that psychopaths are not psychotic, with hallucinations or delusions. Rather, psychopathy is a personality disorder, like borderline or narcissistic personality disorder. It is where aspects of personality are so extreme that they result in problems, for the individual and those around them. It is the nature of those magnified personality traits that is so destructive in psychopaths.

These individuals are callous, lack empathy, are unable to feel guilt or remorse. In their dealings with others, they are shallow, deceitful and highly manipulative, and frequently arrogant to the point of grandiosity. From a young age, they often exhibit antisocial behaviour and aggression that is usually premeditated, although they can also display reactive aggression.

Unsurprisingly, rates of psychopathy in prison populations are very high, with estimates of up to 25 per cent of men and 17 per cent of women in the United States. In the general population it is considerably rarer, with estimates of about 1 per cent of adults having psychopathy.

As with other personality disorders, the origins of psychopathy are thought to be related to a complex interplay between genetic and environmental factors.[53] Again, genetic variants in a range of chemical signalling pathways have been implicated, although those variants identified seem to play a very small role in influencing psychological traits such as callousness or lack of emotion. Familial factors also play an important part: harsh and inconsistent discipline, conflict with parents, low parental warmth and responsivity, and dysfunctional family relationships. Unravelling these genetic and environmental factors is problematic since parents with psychopathic traits will be more likely to parent in a particular way.

The net effect is a blunting of some (but not all) emotional traits, such as fear, empathy and social affiliation, and a reduced ability to detect the fear and distress of others. Indeed, the brains of psychopaths show a reduced response in the amygdala, 'the fear centre', in the face of distress of other people and when it comes to learning from negative emotional stimuli such as fear or pain. Their ability to make moral judgements that involve emotional responses to a particular action is also impaired. As with other personality disorders, structural changes within the brain can also be seen, predominantly in the brain areas involved in learning through fear and processing information regarding reward and punishment, but also more widely within the brain.

In prisoners, the severity of psychopathy correlates most strongly with loss of grey matter in the temporal lobes and limbic system.[54] Additionally, psychopaths also show differences in the connections between various brain regions, particularly those white matter tracts, those highways of signals that connect the pre-frontal cortex to the amygdala, the pathways mediating integration of emotions and decision-making.

These aspects of humanity have relevance for us all, however. Our ability to put ourselves in someone else's position, to empathise on an emotional level, is fundamental to kindness and the curbing of cruelty. Undermining this human trait, of empathy, is core to the ability to inflict violence, to take up arms. The easiest way of achieving this is through dehumanising the enemy, of othering them; we empathise most with those who we see as similar to ourselves. It is one of the challenges of training soldiers – to get them to kill. In several conflicts, the proportion of soldiers aiming shots to miss has been staggeringly high.[55] Hence the use of straw-filled sacks to practise bayoneting, or human-shaped targets without detail, in the specific context of combat, to lessen empathy by removing humanity from the enemy. An attenuation of empathy, akin to psychopathy in very specific circumstances, is crucial to the act of killing.

* * *

Wrath – the emotion of anger and its consequences – clearly has benefits. Without anger we would be unmotivated to strive, to defend ourselves, to defend our group. Both anger and aggression clearly have a neurobiological basis, an underlying circuitry within the brain that influences our propensity for that emotion and our response to it. How our brains have developed is defined by evolution and the genetic factors that are selected for and against to aid our survival. But also by our environment, the world we are brought up in, the families we are part of, the fabric of the society around us. We cannot change our genes, but everything else is potentially within our control.

When I write 'our' control, I mean society rather than us as individuals. For society at large, mechanisms that regulate our behaviour, that lessen emotional or physical violence within the environment that we grow up in, will influence our individual propensity to wrath. Legislation, ethical or religious codes for example: even the use of seatbelts in cars to reduce the risk of traumatic brain injury. These sorts of measures are crucial to our world, to prevent those normal and useful emotions spilling over, unleashing pain, misery, conflict and war, on our families, our societies and our world.

But if there is one thing that people like Tom and Jono illustrate, it is that, for us as individuals, our wrath has a grip that is stronger than our intrinsic personalities, at least for some of us. That our behaviour does not always reflect who we are, our 'soul'. That extrinsic factors – medication, injury or functional disturbance of the brain – rather than our values, can cause us to act in ways that contravene our moral code. However, that dividing line between what constitutes normality and pathology shifts in the sand. That line is blurred by the prevailing winds of our views on morality, legality, philosophy and medicine.

Jono's wife, Hannah, a criminal lawyer, reflects on the people that she has met in her line of work. 'It is one thing I have learnt, just to be patient. I am dealing with people who commit crimes all the time, and also people who have just transgressed once, and suddenly their world is about to collapse. Very rarely,' she says, 'you will meet someone who is just pure evil. Normally, there is a back story there. Mental health. Something traumatic in their lives. Sometimes they have a medical condition. I think we are quick to judge, to think, "Oh, this is just a bad person." Jono is not a bad person, quite the opposite. I know him; I knew him before the drug. It is not him. I think we can all be guilty of assuming who people are. But that is not necessarily the case.'

2.

Gluttony

I thought of the bear, the last of his clan . . . I had intuited his panic, the kind of panic that resulted only from extreme hunger, the kind of hunger that drove you crazy, that drove you south, along highways and through streets, into a place you knew instinctively never to go, where you would be surrounded by creatures who only meant you harm, where you would be going to your own inevitable death. You knew that, and yet you went anyway, because hunger, stopping that hunger, is more important than self-protection; it is more important than life.

Hanya Yanagihara,
To Paradise[1]

The hospital I was working in as a junior doctor was in a very deprived part of London. It was renowned for being hectic, chaotic and relentless, and that night it was living up to its reputation. I was covering general medicine and had been called to see a patient just brought in by ambulance. As I drew back the curtains of the cubicle in the emergency department, I saw a man rising like a mountain from the bed he lay on. The normal hospital gurney that usually sat in the centre of the cubicle had been replaced by an oversized and reinforced bed.

Beneath his bulk, it resembled doll's furniture. James* lay there, helpless as a baby, weighed down by his own sheer magnitude.

In his late thirties, James must have weighed at least a stone for every year of his life. He had stumbled in the bathroom at home and fallen into the shower cubicle. Wedged firmly between the wall and the bathroom enclosure, like a cork in a bottle, too embarrassed to allow his partner to call for help, he had remained there for three days. Eventually, his desperation had overcome his shame, and emergency services had been called. The fire brigade eventually had to destroy the bathroom to free him and remove him from his flat and into the waiting ambulance. It had been his first foray into the outside world for many months.

He lay on the vast hospital bed in front of me, his breath laboured as the weight of his own body squeezed the air out of his lungs and left his heart failing. I clutched a syringe, faced with the impossible task of finding a vein to draw blood from. His skin had disintegrated in places, some related to lying in one position for so long, unable to move, the tissue starved of blood through compression. In other places, this looked much more chronic. As I tried to find venous access, I pushed aside slabs of skin and fat, to find that, in the folds of his flesh, in the dampness of the creases, fungal infection had taken hold, the skin angry and sore.

So began James's long admission, initially to normalise his kidney function, caused by dehydration through his limiting his fluid intake while in his tile and glass coffin, a futile attempt to reduce the need to void his bladder. Then, addressing his heart failure, his diabetes, his blood pressure, his sleep apnoea, all the while trying to treat his broken-down skin. Throughout, he remained on a hospital diet, with normal quantities of food;

---

* Name and some details changed.

yet despite this, his weight refused to shift over the passing weeks.

In snatched conversations with him and his partner, between various procedures and treatments, she exclaimed absolute mystification as to how James had reached the size he had, and they both insisted it was his 'glands'. It was obvious on talking to her that she had a learning disability, and indeed, so did James; neither of them appeared to have a good understanding of healthy eating and lifestyle. James was also troubled by a deep depression, anxiety and agoraphobia. The mystery was solved when his partner was discovered by nursing staff to have been smuggling in vast quantities of food – pizza, chocolate, chips, curry – to supplement James's diet, at his request. All consumed surreptitiously during visiting hours.

The grimness of James's situation was pretty evident. To my deep shame now, his predicament was the subject of dark humour among the junior doctors tasked with looking after him. We were pandering to the widespread view that obesity, the most obvious marker of 'gluttony', was a moral failure, indicative of laziness, a lack of self-control. That our moral value, even our sexual or intellectual value, was somehow a function of our body weight.[2]

In the frenzied battlefield of the hospital, our focus was more on sorting out his physical medical problems rather than seeking any sort of understanding of what had led to this point, which was largely left to the psychiatrists. But even without detailed examination, it was apparent that James had a very pathological relationship with food, a salve for his poor mental health, administered by his partner. And that our appetite is orchestrated by factors beyond the simple need for calories.

While James's predicament illustrates that our emotional relationship with food is a complex one, sometimes a source of comfort or pleasure in the setting of unhappiness or serious psychological distress, there are other reasons for 'gluttony'. These are nothing to do with psychological factors, but originate from the deep recesses of our genes, our guts and our brains. And, as we will see, the clues to these ingredients that define our appetite come from some unexpected places.

* * *

The *Ursus arctos horribilis*, the grizzly bear, must be one of the most successful yoyo dieters of the natural world.[3] In the autumn period, the bears have an almost insatiable appetite, a gluttony unparalleled, consuming 20,000 kcal or more daily, gaining up to 4 kilograms per day. The degree of weight gain is astounding, increasing body mass by up to 50 per cent over the course of a season. The bears do not gorge for the joy of eating, however. This is all for a purpose, for in hibernation they may neither eat nor drink for up to seven months at a time. They survive solely on their fat stores, with even water being a by-product of the metabolic process that breaks down their fat reserves. In the lean, brutal winters of their environment, this accumulation of fat is the only thing that stands between their survival and starvation.

But these morbidly obese bears (at least morbidly obese by human standards) do not seem to suffer any ill consequences. Unlike us humans, they do not exhibit compromise of their breathing, their circulation or their metabolism with this massive weight gain. The grizzly appears to have important physiological adaptations that not only modulate the way that glucose is converted to fat and broken down again, but also how their bodies regulate blood sugar throughout the year. What would cause significant diabetes in humans leaves the morbidly obese grizzly unaffected.

The adjustment of biochemical pathways mediated by insulin,

the hormone that drives the storage of glucose as fat, and glucagon, which has the opposite effect, might explain some of this weight gain and loss. But it is not just a matter of how food energy is stored or burnt off. These bears see a massive surge in appetite that is time-locked to the change of seasons. This increasing dietary intake is driven by changes in at least one specific biological system that regulates appetite.

And this system is not only relevant to the grizzly bear. It has implications for us humans too.

* * *

I meet Alex in her house, about an hour outside London in a small market town. We settle down to chat in the dining room. Alex, twenty-eight years old, is fresh from the hairdresser's this morning, and her hair is coiffured, short at the back and sides – a dense shock of pale blonde hair forming a low, full-bodied fringe that almost touches her glasses.

There are immediate clues that this is no ordinary home, however. The dining room is stripped bare of food and drink, except for two dispensers full of sugar-free cordial, one purple and one orange. The kitchen door in the hallway beyond remains firmly locked, and indeed at the back of the house, the food-waste bin also has a prominent lock on it. The house is shared, and Alex has six housemates. She and all the other residents of the house have the same genetic condition, and require support; two care workers sit in the office behind the room that we are in. The locks on the bin and kitchen door point to this genetic disorder's major manifestation.

'I have PWS [Prader-Willi syndrome],' Alex says, smiling. 'Everyone here has PWS; some have got [type 2] diabetes too.' I ask her what that means. 'It is a genetic problem. It is mainly the food side. It's very hard to control your diet.'[4]

Prader-Willi syndrome is a rare genetic disorder, affecting roughly one person in every 10,000–30,000 people worldwide, regardless of gender or race.[5] And Alex is absolutely correct in that excessive appetite is the most obvious expression of PWS. A gluttony of sorts, but entirely related to her condition, without any moral implications. And this, her unlimited hunger, is only one facet of the disorder.

Kate and Jon, Alex's parents, provide a more complete picture of how PWS has affected Alex. They are both from the United States, Kate from the East Coast, and Jon from the Midwest, but have been living in London now for over thirty years. They were originally transferred here by their company for a limited term of two to four years. I joke that they were captivated by the sunshine. Jon adds, with gentle sarcasm: 'And the good food.'

Alex is the eldest of their four children; the youngest is twenty-two. 'It was the first pregnancy,' Kate tells me. 'So it was hard to know how to benchmark it, but it felt normal. I had an amniocentesis and scans, and everything felt normal. Closer to my due date, they were worried that she looked small, and that she wasn't growing as much as they thought she should.' Jon is six feet eight inches, and Kate is six feet tall, so a big baby would be expected. 'There was also less foetal activity too, and that caused them to do a few more scans. They then decided to induce me on my due date, rather than letting me go beyond term.'[6]

The delivery itself was problematic. The obstetricians had to use both a ventouse – a suction cup attached to the baby's head used to pull the baby out – and subsequently forceps. When Alex was finally born, initially all seemed well, and her Apgar score* was good, but at second assessment she was not crying

---

* The Apgar score is a standardised scoring system for assessing the condition of infants after delivery. It comprises the colour of the infant (i.e., pale, blue or pink), heart rate, reflexes such as grimacing or crying, muscle tone and breathing.

or flailing, and so she was whisked off to the special care baby unit. When Kate tried to breast-feed her, Alex could not suckle, and doctors inserted a feeding tube to give her the nutrients she needed. It took three weeks until she was discharged home – once Kate and Jon had learnt how to use the feeding tube to deliver expressed milk.

The doctors noted Alex to have hypotonia – a floppiness related to low muscle tone – but initially she was just given the generic label of 'failure to thrive', the paediatric term for children whose weight or degree of weight gain falls behind children of a similar age and sex. 'It was stressful,' says Kate. 'She didn't hit the normal milestones; she wasn't doing the things that a normal baby would. She wasn't able to suck. Lifting her head up, crawling, sitting, she did them, but materially later than her peers. I recall going to meet my NCT [National Childbirth Trust*] group, and all the other babies would be sitting up or laughing and smiling.' Alex, she continues, 'was a bit more of a lump, a log. Floppy.'

In those early months, Jon and Kate maintained a degree of optimism. 'You assume your baby will grow out of it. In a way, it being a first child is helpful, right? Because the relative benchmark differential wasn't driven home in quite the same way it would if you had other kids.' And they were given tremendous support by a charity called KIDS, providing help for families who have children with special educational needs and disabilities, with a physiotherapist visiting them at home every week or two. Kate smiles: 'There was one lovely little silver lining, which is kids with PWS are really good sleepers. And they don't cry a lot. So, it is a little bit like God smiling down at you saying: "Well, this parent needs a little bit of a break."'

---

* In the UK, many parents, especially first-time parents, go to classes organised by this charity to gain education about pregnancy, childbirth and parenting skills, in small groups where everyone has roughly the same delivery date.

Alex received a diagnosis remarkably quickly, at around seven months of age, in a highly specialist clinic. The paediatrician noted the features of PWS immediately, and a genetic test confirmed the diagnosis. 'In retrospect, all the telltale signs of Prader-Willi were there,' says Kate. 'But they're all very subtle. Having small hands and feet, having almond-shaped eyes.' Later, when speaking of their experiences going to conferences for children and families with PWS, they tell me that all the children with PWS have similarities: in addition to the eyes and hands, they are often short in stature, wear glasses, and have quite large foreheads, and their gait can be unusual. 'They have a [particular] shape as they start to put on weight. And they are pale and often blonde . . . their skin will be significantly paler [than their parents'].'

The diagnosis was shocking and upsetting for the family. They received a brief description from the clinic, and then ordered in medical textbooks from their local library, it being the pre-internet age. 'The pictures they used [in these textbooks] were as stereotypical as could be,' Jon recalls. 'You get this tome, and it has pictures with the anonymised eyes of a morbidly obese three-year-old. It's not very heart-warming.' They recall the fear and worry, compounded by the thought that all these problems that they were reading about would hit Alex simul- taneously.

Meeting other children with PWS at these conferences was a huge relief. Looking back, they wish it had been made clearer that they would not need to tackle all of Alex's problems at the same time, that they would arise at different points in her life, or not arise at all.

Immediately after the diagnosis, Alex was taken off food supplements, in the knowledge of what was to come. It seems perverse that a child may suddenly go from being unable to put on weight due to difficulties suckling and swallowing, to being unable to stop weight piling on. Until the age of two or three, Kate and Jon found it relatively straightforward to control Alex's access to food, but as she became more mobile, evidence of her excessive appetite became more apparent. Her relationship to food, then and now, is extreme.

'I think she is always thinking about food,' says Kate. 'She always wants to know what the next meal is. "What are we eating? How many appetisers are there? How many people are coming? How many [portions] does that mean for me?" I liken it to being a drug addict. And she just cannot help herself. Her brain thinks it is always hungry.'

Experiments in parabiosis – two lives conjoined

The view that body weight is simply a function of how much you decide to eat and how much you choose to move is one that fuels countless weight loss and exercise programmes. But life, and indeed biology, is rarely so simple. From the 1920s onwards, observations in dogs, monkeys and rats pointed to one particular area of the brain being vital to the regulation of appetite and weight.[7] Damage or tumours of the hypothalamus – a tiny midline area deep in the centre of the brain behind the eyes – was noted to induce rapid weight gain and unleash the appetite of these animals (Figure 4). This observation led some researchers to think that this area of the brain contains nerve centres that normally control food intake. How these clusters of nerve cells could do this was much debated:

It is difficult to see how they can directly measure the number of calories expended in a given period, and then regulate feeding to provide just this amount of energy. Alternatively, the regulating centres may be sensitive to some change in the body which follows the intake or expenditure of energy, and then may inhibit or encourage eating until the changed quantity has been restored to normal. This would be a 'feedback' control system: the controlling centres would be acting on information, fed back to them from the periphery, about the behaviour of the quantity they stabilise.[8]

Rather than these nerve cells measuring how many calories were being burnt, and adjusting appetite accordingly, G. R. Hervey, the author of the above statement and a researcher at the University of Cambridge in the 1950s, considered another explanation more likely. He proposed that the hypothalamus might monitor the overall energy state of the animal, through a 'feeding back' of information from the body back to the brain. In a rather gruesome series of experiments to clarify this feedback control system, Hervey studied rats, which he surgically conjoined. He would take two animals, littermates of the same sex whose weights were similar, then surgically join them by opening up their abdominal cavities and suturing them together. He would then inject a blue dye into the femoral vein of one rat, and confirm that there was an exchange of blood plasma between the two animals by taking a blood sample from the other rat two hours later, ensuring that the dye was also present.

Hervey would then damage the hypothalamus surgically in one of the pair of 'parabiotic' animals (the term parabiosis refers to the joining of two individuals or animals). Those rats with destruction of the hypothalamus would exhibit a dramatic increase in appetite and marked weight gain, to such an extent that they would occasionally choke to death on food, so voracious were they. But the partner rat, with an intact hypothalamus, would eat less, becoming markedly underweight, indeed emaciated. As Hervey remarked: 'Larger amounts of nutrients than normal were presumably circulating in the animals which were overeating and laying down fat, of which some portion would have been carried by the cross-circulation to the partners; the partners would therefore have been expected to gain weight if they had not eaten less.' He argued that the partners of animals with hypothalamic damage became thin because their own functioning hypothalamus reduced their intake of food. Essentially, the overfed body of one animal produced some sort of signal to influence the intact hypothalamus of the partner, to suppress feeding in the context of an intact brain. 'On this interpretation, therefore, the experiments demonstrate what in a normal single animal would constitute a feedback system.'[9] Hervey's studies were supportive of the idea that feeding behaviour is adjusted by the amount of stored fat in the body, and that areas of the hypothalamus are sensitive to a chemical derived from stored fat.

The search for the hunger chemical

What this chemical that signals fat storage might be remained a mystery for a number of years, but insights came from other rodents, this time the house mouse. Researchers breeding mice found that some of the offspring produced by one such breeding stock, in the summer of 1949, were almost indistinguishable from others in the same litter, at least until four to six weeks of age. Over the next few weeks, however, these mice would pile on weight, fuelled by an excessive appetite. By three months they would weigh twice as much as their peers.[10] Two black and white photos from that original paper show the normal and obese mouse side by side. On the left, the two mice are pictured at twenty-one days of age, already one slightly rounder and broader. But by ten months, the difference is staggering: one mouse inflated to such an extent that its pointy snout and small ears sink into the fat, a circular furry blob with a tail attached, weighing over three times more than its normal littermate.

Of 212 mice born to the parents, 43 were obese. This ratio implied that the apparently normal parents were each carrying a single copy of a defective gene predisposing to obesity but masked by the presence of a normal copy of the gene,* and that one in four of their offspring would bear two copies of that mutated gene (these obese mice were referred to as *ob/ob*, carrying two mutated genes). Subsequently, studies in the 1970s, utilising the technique of parabiosis – the conjoining of mice surgically that Hervey used – suggested that these genetically obese mice lacked a blood-borne factor that regulates body weight.

Identifying this mysterious substance took decades of further painstaking genetic work. In 1994, the mutated gene in mice, and its equivalent in humans, was fully identified. This facilitated the discovery of leptin, the protein produced by the *ob* gene. It is this chemical that is secreted by the fat cells themselves, signalling directly to the hypothalamus the levels of fat storage within the body. The brain was monitoring levels of fat stores within the body by sampling levels of leptin within the bloodstream.

Thus, these experiments clearly refuted the idea that the weight of humans and mice alike was simply a measure of willpower,

---

* This ratio of one in four offspring being affected by a disease when both parents are healthy is rather characteristic of a recessive genetic model of inheritance, where two copies of an abnormal gene are required in order to develop the disease. If both parents carry one dominant and one recessive gene, then one quarter of offspring will inherit two copies of the dominant gene, half will have one dominant and one recessive gene (and will therefore not outwardly manifest, but will be carriers), and one quarter will carry two copies of the recessive gene. Those carrying these two recessive genes will be the only ones with the disease. This ratio may of course be influenced if the disease increases the risk of foetal death prior to birth.

an indication of sinful gluttony or virtuous temperance. Instead, they confirmed that feeding behaviour and body weight are determined by a well-characterised physiological system. They also demonstrated that adipose tissue – fat – is not just a passive store of calories, but is actually active, signalling to other parts of the body, and involved in its own regulation.

This leptin system is therefore fundamental to controlling appetite – in rats, mice and humans, but other species too. Nevertheless, if some humans are thin and others are fat in the same environment, with the same relatively unrestricted access to calories, this implies that the setting of the leptin 'thermostat' may differ between individuals, that the tipping point for feeding behaviour from eating more to eating less is at different levels of fat stores. Indeed, the grizzly bear also tells us this: that even in the same animal, the sensitivity of the hypothalamus can be adjusted. As the bears reach autumn, their brains become insensitive to leptin, and they never achieve satiety. But as they reach the point at which they begin to hibernate, this insensitivity reverses, and the effects of leptin on the brain suddenly kick in, thus suppressing appetite, in preparation for the long winter ahead.

Leptin and the hypothalamus have some relevance too for people with conditions like PWS – for Alex and her never-sated hunger. Her appetite is a product of this hunger thermostat never turning off, set to maximum in perpetuity.

Weight is not all about leptin

For all of us, our appetite is not just regulated by the *ob* gene and leptin, however. There are many other genes in the mix. Some 40–70 per cent of each individual's predisposition to obesity is inherited, although most of these genetic contributions currently remain unknown.[11]

Even the study of families with extreme obesity due to a single genetic mutation tells us that leptin is not the only genetic factor at play. Human parallels of the *ob/ob* mouse, where individuals carry two abnormal copies of the gene and exhibit disease, are extremely rare. These genetic mutations have only been described in a small number of individuals in the world – two families in Pakistan, and one in Turkey.[12] Like the leptin-deficient mice, all these children were born at a normal weight, but within the first three months of life showed eye-popping weight gain, tipping the scales at more than 20 kg by one year old, and more than 50 kg at five years old.

Another few humans with mutations in the receptor for leptin, hampering the detection of leptin rather than the production of it, have also been found. In these people, no matter how much leptin is produced by body fat, their brains simply do not detect it. Whether it is the lack of leptin, or the inability to sense it, the effect is the same: the uncontrolled weight gain associated with limitless hunger. In yet other families, mutations in genes that are fundamental to the normal function of those leptin-sensing neurones in the hypothalamus* have all been identified as causing early marked obesity.[13]

For the majority of these genes, it is the inheritance of two copies of the mutated gene – one from each parent – that results in disease, like the recessive *ob/ob* mouse. But there is some evidence that even carrying one copy of a mutated gene, previously not thought to produce any ill-effects, may predispose to having a higher body-mass index (BMI) and body-fat mass. This suggests that these genes may have a broader role in the hereditary contributors to obesity.

For most of us, however, the genetic determinants of our

---

*   These leptin-producing neurones reside in the paraventricular nucleus (PVN) of the hypothalamus. Those genes identified in other families, called *POMC*, *PC1/3*, and *MC4R*, are crucial to normal functioning of leptin signalling.

appetite and weight are a little more complicated. Rather than having rare genetic mutations, it is normal variants of genes that we carry that influence our weight. In a huge study of almost 340,000 individuals, 97 genes associated with BMI were identified.[14] Up to 21 per cent of variation in human BMI can be accounted for by common identified genetic variants (also termed polymorphisms) within these genes. To some extent, it is the combination of variants of these ninety-seven genes that you inherit from your mother and father, that define your body weight.

The distinction between a mutation, like that in the leptin gene, and a common genetic variant is an important one. Mutations are rare changes in the genetic code that often result in dramatic changes in the functions of our genes to the extent that they almost invariably cause disease. In contrast, these common variants within our genes do not have such devastating effects so that their presence guarantees a disease or disorder. Each variant conveys a more minor effect, a more subtle change to our biology. It is the summation of all these minor variations in our genes, and their more delicate consequences, that give rise to a particular trait or contribute to the risk of a disease.

Many of these common variants associated with obesity are in genes that influence the function of the hypothalamus. They shape the setting of the hunger thermostat, and at what level of body weight we stop eating.

What benefits do these common genes confer?

For genetic disorders arising from rare mutations, like those few families with leptin or receptor abnormalities, it is obvious why they are so infrequent. Many of the individuals with these rare genetic causes of obesity are infertile, and so these mutations are to some extent maintained in the human population through new mutations, not reproduction. These mutations arise by chance, a randomness of the universe, an error written into the genetic code during the replication of a cell.

For the rest of humanity, however, where obesity is so heavily influenced by common genetic variants rather than these rare and highly damaging mutations, one obvious question arises. For any genetic variant that is widespread in the population, there must be some sort of evolutionary process driving the spread of that version of a gene through the generations. If these variants had distinctly negative effects, influencing survival, over time they would have been selected against, becoming rarer rather than more common. This question is not only specific to gluttony but, as we will see, to the other 'sins' too, and indeed any human trait under genetic influence.

At some point, these common variants must have arisen as a form of mutation, a mistake in the copying of the DNA sequence during the production of sperm or an egg that gave rise to a single individual. If that mutation was damaging enough to cause serious disease, then the carrier of that mutation or its offspring would have been affected severely enough to impact their ability to pass on their genes, to limit their survival.* Hence these damaging mutations would remain rare in subsequent populations, like those humans with a leptin gene mutation. But if that alteration in genetic sequence is less likely to cause disease, and indeed may contribute to some benefit for survival, then carriers of that mutation would be more likely to pass these on to their offspring. Over time, these mutations would

---

* The exception to this would be mutations that clearly cause disease, but which do not affect individuals before they have had a chance to procreate. An example of this is the BRCA1 gene, which predisposes to breast cancer. Women who carry the BRCA1 gene have a 55–72 per cent of developing breast cancer by their 70s–80s. This gene mutation is present in about one woman in 400–500 – still rare, presumably because there is no particular survival advantage to carrying it.

no longer be rare, and would spread through the generations to become common variants.

Presumably, therefore, at some point in human evolution a genetic variant that confers survival advantage has arisen through chance alone and then proliferated due to evolutionary pressures.

With regard to obesity, this variant must have been protective in some way, predisposing to survival, to being passed on, yet also inclining us to gain weight. Hominids first appeared some 5 million years ago, developed an agrarian society about 10,000 years ago, and began to industrialise less than 200 years ago. Therefore, almost 5 million years of evolution might favour the dietary conditions of our hunter-gatherer and agrarian ancestors, rather than the last hundred or so years that have seen obesity become an epidemic in human society.

The thrifty gene hypothesis

Evidence therefore points to a major role for our genes defining our body weight. In view of current levels of obesity, these genetic variants that predispose to obesity must be very widespread indeed in the human population. For this to be the case, there must be a reason for evolution to favour such genetic variants – some survival advantage that drives the holders of these genes to survive and procreate more, to create more offspring who inherit this advantage too.

If I were evolution – or God, depending on your inclination – I would certainly promote genes that would either drive us to store as much fat and as many calories as possible at times of plenty, or to use food energy as efficiently as possible. These genes would represent an overdraft facility in times of famine, a cushion against periods of starvation. Yet those same genes would drive obesity, and associated conditions like diabetes, high blood pressure and other obesity-related ill health, and thus lower survival advantage in times of feast. A genetic inheritance that drives metabolic efficiency would have conferred a major advantage to survival in our ancestors, to gain weight more easily and survive famine. In our current world of plenty, which is but a brief moment in time in the course of human evolution, this gene (or genes) now drives obesity, and diabetes. Perhaps then, it is the inheritance of genes that promote this frugality of energy – 'thrifty genes' – that is responsible for our current obesity epidemic?[15]

While this makes total sense, evidence for the existence of such a thrifty gene is scant, and major criticisms have been levelled at this hypothesis.[16] Anthropological evidence suggests that many of the populations currently demonstrating very high levels of obesity and diabetes, like Pacific Islanders, do not appear to have a history of famine or starvation; tropical islands surrounded by fish-filled warm seas tend to have a stable abundance of accessible calories. This is, however, predicated on the possibility that these populations have had sufficient evolutionary time since the migration of humans out of Africa for their genes to adapt to their new surroundings.

A second criticism is that modern-day hunter-gatherers would be expected to gain weight easily between periods of famine, something that is not borne out in reality. It may of course be that these genes only permit very limited weight gain in the context of a pre-industrialised diet, and that this is only supercharged in our current post-industrial societies with their never-ending supply of tasty high-calorie food.

Famine in and of itself may not be enough of a selective pressure for evolution to explain obesity rates of 20–30 per cent in today's societies though. Lethal famine was rare in pre-agrarian societies due to the diversity of food sources, and was only an issue when climate decimated crops – more likely in the agrarian age. If we then consider that famine was only a more recent pressure in our evolutionary path, since we became dependent on agriculture, some 400–600 generations ago, then this is too short a time for these genes to be so widely dispersed throughout human populations.

If not thrifty genes, then what else?

If there are doubts about the 'thrifty gene' hypothesis explaining the spread of these genetic variants among the human population, what might be other drivers of our inheritance of obesity?

Perhaps the major evolutionary influences on body mass in our ancestors were not only famine – not being too lean to survive food shortage – but also predation – not being too fat to run away from predators that would eat us.[17] In light of the argument about whether famine is such a major evolutionary pressure, some scientists have argued that it is predation that is more important, preventing us from becoming too large. Since we humans have become better masters of our own destiny, with weapons and fire to see off our predators, essentially this limitation on our size has been removed. In possession of sticks and spears, our survival has become less dependent on our ability to run away quickly from predators, and being large or fat is less of a disadvantage. Larger individuals, those who would previously have perished in the jaws of wild beasts, now survive and procreate more. As a result, these genes have gradually percolated more widely throughout human populations.

Scientists point to the change in body weight in hominids over time; 2–4 million years ago, the *Australopithecus*, our early ancestors residing in the Sahara, preyed upon by *Dinofelis* sabre-toothed cats, weighed 29–52 kg; *Homo erectus* acquired weapons and fire at around the same time as the cats became extinct, around 1.4 million years ago, and grew significantly larger in body size.

There is another possibility, however, nothing to do with storing calories for time of famine, or running away from sabre-toothed cats. In essence, that the genes that predispose us to obesity are not selected based upon our body weight, but that this is simply a bystander effect of our ability to survive in the cold. It stems from the observation that, even accounting for environment, education and economic factors, obesity rates are hugely variable in different parts of the world, as are the genes that have been linked to obesity. In fact, crucial to this theory, many of the genes associated with obesity, and obesity-associated problems like type 2 diabetes, have also been implicated in physiological adaptation to climate. In contrast to previous theories, it is climate that has selected for genes that determine metabolic rate and how much fat we burn.[18] Proponents of this view, termed the 'maladaptation hypothesis', point to numerous observations to support their theory: that Scandinavian countries have much lower obesity rates than other European countries, despite having similar diets; that Caucasians and East Asians have lower rates of obesity than African Americans; that Inuit have a higher metabolic rate than African Americans. Basically, the colder the environment your ancestors adapted for, the faster you burn fat, and the more heat you produce. One genetic example is a gene called *UCP1*, active in the visceral fat termed brown adipose tissue, or BAT. The product of *UCP1* is a key factor in the breakdown of BAT to produce body heat. Variants within this gene show associations with significantly different levels of body fat.[19]

For Alex, too, the source of her irrepressible hunger and risk of weight gain is in her genes. Unlike the majority of us, though, it is not related to these common genetic variants – be they thrifty, related to the lack of predators or due to burning energy to survive in a cold climate. Alex's condition is not due to a single genetic mutation either. Prader-Willi syndrome is clinically and genetically more complex.

In additional to its impact on appetite, PWS affects stature, body shape and muscle tone, as we have already seen. One of the other features is intellectual disability. 'Most adults with PWS have the emotional maturity of an eight- to ten-year old,' Kate, Alex's mother, says.

While Alex is an eloquent young woman, getting a feel for her own insights into her relationship to food is difficult. On one level, she clearly understands the need to regulate her diet. 'If you don't control your food, you can easily get very big. Without weight management, if you get very big, you might die very quickly with PWS,' she tells me. When I ask her if she is always hungry, she says: 'Not all the time, no. Not after a big meal.' Later, however, she describes how at one point when she was living a little more independently, each meal would be placed in the fridge shortly before mealtimes. I ask her what would have happened if all three meals had been put in the fridge at the same time. 'I probably would have eaten it all in one go.' I suspect her definition of a big meal, the volume of food required to sate her hunger, is very different from most people's. Her parents say that she understands the need to regulate food in theory, but her actions do not marry with that understanding: she simply cannot control her food intake. It is also difficult to grasp what it is that Alex actually feels: whether she feels the hunger that most of us sense, but it is simply never sated, or whether the intensity of the hunger is supercharged, beyond what other people would normally experience.

In PWS, there is a double whammy. Not only is the appetite excessive, but due to the muscle issues and reduced mobility, individuals with PWS actually have a lower calorie requirement than normal. There are other risks of an insatiable hunger too. 'People with PWS don't have a gag reflex,' Kate and Jon explain. Because of the impaired gag reflex, overeating and regurgitation will occasionally lead to people with PWS choking to death. Due to distension from vast quantities of food, perforation of the stomach may also be fatal. Occasionally, eating food not for human consumption, or even non-edible substances may cause significant harm. Alex's parents report that she has eaten dry cat food before, and that they have heard of other people with PWS eating scraps of carpet, for example.

An extreme appetite necessitates extreme measures. Jon and Kate have gone to enormous lengths to control Alex's food intake. They have never had chocolate or sweets in the house, but Alex says she is, in any case, indiscriminate about her food selection – when I ask her if she has a preference for sweets or cakes, she simply replies: 'Any food!' Her parents focus on fruit and vegetables, any way of providing bulk with fewer calories. A birthday may be marked by a clementine with a candle stuck into it, and early on they found a wind-up wooden birthday cake with candles, that Alex took great delight in.

But these sorts of methods are simply not enough. They describe Alex as having a 'foraging instinct' that they are compelled to curtail. I can see the vestiges of locks on the bin and the cupboards and drawers of the kitchen in the family home. In recent years, they have simply taken to locking the kitchen door when Alex comes to visit. They describe family meals or trips out as sometimes being points of strain, in part due to Alex's needs but also due to other people too. 'As a parent you have to be a vigilant policeperson. People might be well-meaning – "Oh, that little girl is so adorable. Can she have a lollipop?" And you feel quite the grumps. "No, actually she can't have a lollipop." Often, I would just pretend she was diabetic.' Kate describes an event in a Paris art museum when

Alex was becoming distressed, due to hunger, and someone asking Kate why she could not keep her child quiet. She responded assertively and recalls her own fury. There have been many occasions in a shop or restaurant, when Alex has been pleading for food or an extra serving, where her mother has felt the disapproving glare of cashiers or waitresses at their 'overly strict' parenting. (She says she can sometimes see the upsides of having a child with a more apparent disability, for whom the community are more likely to be more sympathetic and supportive.)

Alex will often focus on 'getting her fair share'. Kate tells me that she will often say she is hungry, that it is her snack time, and ask when they are eating next. Food needs to be served away from the table, as otherwise controlling Alex's portion size is impossible. 'She will grab the spatula and say: "I want more." If there are three main dishes on offer in a big family buffet, she will want not a small portion of all three, but a full portion of all three, and we have to intervene. Some family members will say give in and others say don't give in, and that will lead to tension.'

I ask her parents if they think Alex ever feels shamed – the reflected judgement of society and its views on overeating. Kate says: 'That's an area where perhaps a lower IQ helps. I am not sure her maturity would understand shame. There is a much simpler understanding of life, of emotions. And I think that's a benefit for her.' I ask Alex if she feels a sense of injustice, that her food is regulated as tightly as it is. Despite her telling me that she does not, that she understands the need to limit her food, I get the impression from her parents that she does occasionally express that it is unfair. But the primary feeling is simply of hunger.

Kate and Jon describe many examples of Alex's ingenuity in obtaining food. Even now, in anticipation of a visit to her parents' house, a delivery of food from a supermarket will arrive at the door, snacks and other food ordered by Alex online. This tactic is usually thwarted by her parents. In the house where she now resides, where all Alex's housemates also have PWS, care workers ask for any parcels delivered to be opened in front of them to make sure they contain no food, and any purchases when the residents go out need to be accompanied by receipts. In part this is due to requirements to safeguard vulnerable adults from financial abuse, but also to ensure no food has been surreptitiously bought. Alex tells me they are permitted one chocolate bar a week, and one drink with sugar in it at the weekend. All the residents have individualised diets based upon their weight and calculated requirements.

Kate and Jon's vigilance, as well as that of the home, has clearly had benefits. Alex is now not at an unhealthy weight. The impact of lessening control over her diet is apparent, however. Jon shows me a chart of her weight, largely stable for the last few years, hovering between 75 and 80 kg. When she finished school, and went to a college, for the first two years her dietary control was less stringent, and over that period her weight rocketed, reaching a peak of 110 kg. It illustrates the necessity of all of this.

My own experience of parents with children with rare genetic disorders, both professionally and in my own extended family, is that they live in fear of having further children affected by the same disease. But Prader-Willi, despite having a genetic basis, is not usually inherited. It is atypical among genetic disorders in that it depends upon a genetic phenomenon called imprinting.*

---

* Specific regulation of genes through imprinting, based on current evidence, is almost unique to placental mammals and marsupials in the animal world. The reasons for the evolution of this genetic mechanism remain unclear, but the

Usually, both the genes inherited from your mother or father are actively utilised, but there is a subset of genes inherited from one parent or the other, about a hundred in total, that are silenced. It is this phenomenon of gene silencing that the term imprinting describes.[21]

In the process of cell division and replication, the genetic code needs to be copied. There are regions of the human genome, however, that are inherently unstable, more liable to slip-ups in this duplication process. One such region, a small stretch of DNA sequence in chromosome 15 containing several genes, is sometimes deleted during the process of cell division and replication. If this deletion occurs during the production of eggs or sperm, then any embryo resulting from conception will only have one copy of the genes in this region, rather than the normal two copies. Some people carry this deletion and are entirely unaffected. However, the cluster of genes within this region of the genetic code is subject to imprinting. Those genes inherited from one's mother, carried by the egg, are silenced. Thus, if the sperm contributing to a pregnancy has these genes deleted, and the mirror genes in the egg containing the maternal genetic code are silenced through imprinting, then essentially the resulting foetus will have none of these genes functioning. This genetic abnormality is the commonest cause of PWS, responsible for about 70 per cent of cases. This deletion of the region of the paternal chromosome 15 is a chance happening,

---

most widely accepted theory is termed the 'parental conflict hypothesis'.[20] This proposes that imprinting results from different interests of each parent in terms of the evolutionary fitness of their genes. From a maternal perspective, a mother's evolutionary drive is to ensure that, while providing adequate resources to the child she is carrying is important, it should not be at the expense of her own survival, or of future children she might bear. From a paternal viewpoint, the priority must simply be to push their offspring to have as many resources as possible to ensure their child's survival, even at the expense of the mother. Paternally expressed genes tend to promote growth, whereas maternally expressed genes tend to limit it. Imprinting has been described as a genetic 'parental tug-of-war'.

during the formation of the sperm, and therefore most parents with a child with PWS do not have an elevated risk of having another child with the condition.

The precise mechanism for how these genetic changes in Prader-Willi cause such a marked increase in appetite is not fully understood. This chromosomal deletion affects nine genes within this region, and so the exact attribution of particular functions or dysfunctions to individual genes is difficult to ascertain. In PWS, there are a range of hormonal changes affecting metabolism and satiety, including leptin. Additionally, however, people with PWS demonstrate changes in several brain areas, including the hypothalamus, where the hunger centre resides. Those genes silenced in PWS appear to have a role in the normal development of the hypothalamus.* Brain scans have shown that food causes a greater activation of the hypothalamus and the limbic system in people with PWS, fuelling appetite and the emotional response to food, and lower activation than normal in those regions of the frontal lobes like the medial pre-frontal cortex involved in inhibition – in Alex's case the inhibition of eating behaviour.[23]

Kate and Jon explain that when they learnt of the genetic basis of Alex's PWS, and that it was not hereditary, they decided

---

* In some of the broader, more complex genetic disorders where weight gain and intense overeating are part of a clinical picture – those that often include intellectual difficulties and anatomical abnormalities – there is sometimes a common link, although thus far this does not seem to apply to PWS. Their causative genes are linked to a specific aspect of the human cell.[22] The primary cilium, a small hairlike structure projecting from the cell surface, is a feature of almost all nucleated cells. Some cilia move, but these primary cilia appear to have an important role in the detection of chemicals and mechanical stimuli, as well as signalling within the cell itself. And it seems that those cilia in the hypothalamus, in particular, are key to transmitting information crucial to the regulation of feeding behaviour. Simply destroying these primary cilia in the hypothalamus alone, using genetic techniques, causes obesity. What these mutations do is not only affect the primary cilia function here in the hypothalamus, but also throughout the body, hence causing the varied anatomical and intellectual manifestations of these conditions.

to have more children. 'We have met many parents who have a child with special needs, who then are so stuck, so stricken with grief, that they go on not to have any more kids, or even cease to be a couple.' Additionally, they were starkly aware that at some point they would not be around anymore, and Alex would be left to fend for herself. When Alex's siblings were much younger, they were incredibly supportive, and were her friends. They also provided role models, encouraging Alex to learn to crawl and walk. In teenage years, her siblings expressed some embarrassment at her behaviour, and frustration at the lack of treats available to them. 'We would indulge our kids when Alex wasn't around or when we were out,' says Kate. 'And there were moments of pride when Alex's younger sisters stood up at their school assemblies and talked about having a sister with special needs and raised money for a charity with a cake sale.' Jon chuckles: 'Ironic, but also quite nice.' Now the siblings are all in their twenties, and still very supportive. 'If Alex is about to blow, they know how to distract, defuse, and can help with the load sharing,' says Jon. Kate adds: 'I do think it has made them more mature. It has made them better people. Much more emotionally aware and generous.'

Is obesity a communicable disease?

For most of us, a single gene mutation, or even the deletion of a few genes as in Alex's case, does not explain our tendency to overeat and put on weight. A myriad common genetic variants seems to explain 40–70 per cent of obesity, but that leaves 30–60 per cent unaccounted for. After all, even genetically identical twins may have different body weights, even more so if they grow up apart.[24] And the huge increase in obesity levels over only a few generations, without any big change in our genetic make-up, also clearly shows the importance of environment, not just genetics.

Some of those environmental factors may be somewhat surprising. In a Harvard study from 2007, researchers followed a cohort of 12,000 individuals, assessing them repeatedly between 1971 and 2003. They tracked weight gain and obesity over this thirty-two-year period, and mapped changes in body weight according to the relationships between these individuals.[25] What they found was rather amazing: that obesity spreads via social connections, and that the pattern of that spread is dependent on the nature of those social ties. While geographical proximity, i.e. being neighbours, did not seem to be a factor, being a friend or a sibling of someone obese was a much stronger influence on the development of obesity in oneself. In fact, spousal obesity was a much weaker factor than having an obese friend, especially one of the same gender, implying that this was not necessarily down to sharing the same physical environment. Instead, it suggests that people are more strongly influenced by those they feel resemble themselves.

The researchers also concluded that these patterns of spread of obesity could not be explained by the possible tendency for obese people to form ties between each other, nor was it the case that friends simultaneously became obese due to exposure to some undetermined factor. It is not the case that being obese makes you seek out obese friends. Instead, having a close friend who is obese subsequently increases your risk of becoming obese over time.

The authors of this study speculated that this contagion of obesity, essentially acting as a communicable disease, was being mediated by psychosocial factors. Their results implied that one of the most important factors in this contagion was seeing people that we identify with becoming obese. This seems much more important than seeing obese people in our neighbourhood.

The researchers argued that this spread in obesity may have more to do with perceptions of norms among people that we view ourselves as being similar to, rather than direct effects on eating behaviour.

The concept of obesity acting as a communicable disease is an odd one, out of kilter with most of our views on weight – a simplistic assumption that it comes down to energy in versus energy out, and is down to the individual's appetite, diet and physical activity. It clearly demonstrates the complexities of this idea of the 'sin of gluttony', of which obesity is the visible signature, a flag of shame and moral weakness. But if the assumptions made by the authors of this study are correct, then it does perhaps show that acceptability of obesity may be a factor in its spread to others. By implication, making it unacceptable, in moral or any other terms, reduces the likelihood of that contagion.

It may not just be these psychosocial aspects that lead obesity to act like an infectious disease, though. Obesity may in part actually *be* an infectious disease. It may be an epidemic in the truest sense.[26]

A number of viruses have been found in animals to be obesogenic – to drive weight gain – even in the presence of no change in food intake. One virus in particular, an avian adenovirus called SMAM-1, has been found to cause an increased accumulation of body fat in infected chickens, without an associated increase in appetite. When researchers in Mumbai assessed obese human individuals for antibodies to this virus, those positive for antibodies (implying previous infection) were heavier and had a higher BMI than those who were negative.[27]

Other viruses have also been implicated. Antibodies directed at Ad-36, a human adenovirus, were assessed in 502 obese and non-obese residents of the United States.[28] Independent of age or sex, Ad-36 antibodies were found in 30 per cent of obese individuals and only 11 per cent of non-obese individuals. The presence of the antibody, a marker of previous infection with the virus, was associated with a sizeable difference in levels of obesity. In twins, where one was positive for antibodies and the other negative, the positive twin was significantly heavier and fatter than the negative one.

If viruses can trigger weight gain, how do they do this? Many of the viruses that cause obesity in animals infect the central nervous system, but no clear explanation for why they may drive obesity has been found thus far. Ad-36 appears to be able to directly infect the fat cells themselves, and the virus seems to stimulate the rapid development of fat cells and the accumulation of triglycerides (the major constituents of body fat). Furthermore, in vitro studies – in test tubes or petri dishes rather than in living organisms – have found that Ad-36 and some other adenoviruses reduce the production of leptin by fat cells, by as much as five-fold.[29] This virus therefore might have a double effect, enhancing the development and growth of fat cells, while simultaneously increasing appetite by suppressing leptin production.

Micro-organisms and weight gain

That micro-organisms, through infection of our fat cells, may influence our body weight is startling enough. In recent years, however, an even more unlikely mechanism for how microbes may drive obesity has come to light.

In my very first post as a junior doctor, I worked in a busy hospital in northwest London, on the general surgery wards. The hospital was a brutalist concrete monstrosity, ugly as sin, most remarkable for two unrelated aspects. The first was the presence of a paternoster lift, which allowed us to move through the several floors of the building rapidly and efficiently, stepping into the void of the lift shaft as platforms continuously rotated up and down. It was rumoured that patients had died in the lift, wheelchairs falling into the shaft and being crushed by the unrelenting machine. The second aspect was that this busy general hospital occupied the same site as another institution, unappealingly named St Mark's Hospital for Fistulas and other Diseases of the Anus and Rectum. Now housed in a similar concrete carbuncle, St Mark's had been founded in 1835 and was originally called 'The Benevolent Dispensary for the Relief of the Poor Afflicted with Fistula and other Diseases of the Rectum'. The move to its new site had only happened three years before I started my post there, and what happened in the neighbouring building was still a mystery to most of us doctors working in the main hospital.

In the metaphorical bowels of St Mark's, it was whispered that all sorts of unusual practices were taking place: procedures such as porridge enemas; toilet bowls with internal video cameras to visualise and record patients evacuating into the pan. Most stomach-churning were reports of faecal transplants – taking the stools of healthy people, liquidising them in a blender, then instilling the poo milkshake into a patient with a range of colonic diseases, either through an enema or down a nasogastric tube. Whether any of this was true, or just constituted urban legend within the confines of the hospital, these stories made for light entertainment in the doctors' mess in the middle of the night. But the reality of faecal transplantation is unarguable, and its origins are old.

The advent of antibiotics, and more widespread use of these drugs in the aftermath of the Second World War, prompted discussions in academic and clinical circles of the possibility of life in the absence of germs, that these drugs might essentially eliminate infectious diseases as any consequence to human existence.[30] Somewhat naive in retrospect, given what we know of antibiotic resistance these days. Arguments raged between prominent scientists, physicians and Nobel Prize winners about whether the prospect of germ-free humans might liberate us from the historical shackles of fear of infection, or would actually weaken our immune system, placing us at mortal risk of meeting even a single microbe. In this intellectual environment, it was natural for researchers to look at the effects of living in a sterile environment, and from the 1940s onwards, scientists began to develop animals that were free of all microbes.

It was in these germ-free rats and chicks that, in the mid-1960s, researchers found that the presence of microbes in the gut interfered with the absorption of fats. This observation was one of the first indications that gut microbes – the microbiome (strictly speaking, the micro-organisms in our guts should actually be referred to as gut microbiota) – might be associated with weight regulation. In fact, subsequent studies in animals and humans have shown different proportions of types of bacteria present in stool are associated with obesity.*[31]

I remember first hearing about these studies on poo, and those lifeforms that live within us, and being deeply sceptical. As slews of studies correlating our microbiota with various aspects of our health began to be published, it was obvious to me that the make-up of the bugs in our guts was likely to be influenced by a range of factors, such as what we eat, our lifestyle and our own genetics. I thought it likely that it was our diets that were causally related to our health, and that our microbiota were simply a product of our diets too. I was suspicious of many of the claims regarding prebiotics, probiotics, and so on, that they were simply the product of over-eager marketing departments.

In part, I was correct. We now know that our microbiota are

---

* Specifically, the ratio between two groups of bacteria, Bacteroidetes and Firmicutes, differed significantly between obese and lean people.

strongly influenced by a range of chemicals produced by our guts, and other factors, like how we are born.[32] Newborn babies will have very different microbiota depending on whether they are delivered vaginally or via caesarean section. Diet is fundamental: breast versus formula milk, prebiotics, dietary fibre, complex versus simple carbohydrates, different forms of fat, antibiotics – these factors all influence the specifics of the bacteria residing in our gastro-intestinal tracts.

But I was also very wrong. In hindsight, that the roughly 10–100 trillion organisms living inside us (compared to the 30 trillion cells constituting our own bodies) might have some direct influence on our physiology should not have been a surprise to me. In fact, changing our gut bacteria to treat certain diseases formed part of medical practice centuries before we even knew what bacteria were. The earliest descriptions of faecal matter transplantation (FMT), akin to what was being undertaken in our own hospital, date back to the fourth century, where physicians used faecal suspensions, described as 'yellow soup', given by mouth to treat severe cases of food poisoning or diarrhoea.[33] Since that time, FMT has been used in modern medicine too, for the treatment of inflammation of the colon caused by a particularly nasty organism called *Clostridium difficile*, autoimmune conditions of the gut such as ulcerative colitis, and experimentally for conditions ranging from multiple sclerosis to blood disorders.[34] Rather than feeding people 'yellow soup', modern administration is rather more palatable (and less horrifying) – by colonoscopy, enema, nasogastric tube or capsules.

The rationale for our gut microbiota influencing infectious or inflammatory diseases of the gut is entirely intuitive. We have long known that there are good bacteria on our skin that protect us from disease-causing bacteria. We have even used these principles for millennia when it comes to feeding ourselves – many foods are fermented by 'friendly bacteria' in an effort to crowd out harmful microbes. However, the nature of how bacteria in our guts might influence our weight and the constitution of our bodies is a little more puzzling. And it seems that it is not simply the case that our diets influence both our weight and our microbiota, as I had initially presumed.

It is becoming increasingly clear that manipulating the gut organisms residing within fundamentally influences how we process food and regulate energy. A seminal study in 2004 demonstrated that mice bred in a germ-free environment, with sterile guts, gained less body weight and had a lower proportion of body fat than normal mice.[35] Furthermore, when gut microbes were introduced to these mice, not only did they store more fat, but their glucose, insulin and leptin levels all went up. Other studies, again in mice, have shown that low levels of antibiotics in early life predispose to weight gain and increased fat content, especially if the antibiotics are given from birth. If the gut microbiota are transferred from these antibiotic-treated mice to germ-free mice, this has a similar effect on body weight in those recipient mice, leading them to put on weight, strongly supporting a causal role rather than an associative relationship between microbiota and obesity.[36]

The detail of how bacteria in our guts might directly contribute to our energy metabolism remains poorly understood, but it is clear that there are direct actions through altering the amount of calories available for the gut to absorb, as well as indirect actions, through the effects of compounds or chemicals produced by these microbes that influence our own metabolic pathways.*[37] The chemical products secreted by some

---

* One such indirect action is through modulation of a group of chemicals called short-chain fatty acids (SCFAs). Certain gut bacteria produce SCFAs through fermentation of carbohydrates, and these compounds have a range of effects – some local, such as regulation of gut acidity, and production of mucus, and some systemic; when SFCAs are absorbed by the gut, they decrease the production of glucose and fats by the liver.

of these bacteria also have the direct effect of stimulating the secretion of certain hormones by the colon, called PYY and GLP1, which suppress appetite (the rash of new drugs to facilitate weight loss, like Wegovy and Ozempic, are GLP1-like substances, mimicking the effects of GLP1 on the brain and beyond).* Giving these bacterial products to rodents promotes weight loss, although data in humans remains much more limited.

So, if our gut flora do indeed have such an important role in regulating our appetite, our energy metabolism and our weight, could we simply perform faecal transplantation from lean to obese individuals to treat obesity? Unfortunately, results from human studies have been mixed so far. While some researchers have reported improvements in the response of our bodies to insulin, a consistent reduction in BMI has not been demonstrated. The issue with humans, as opposed to mice in a laboratory setting, is that dietary habits are more difficult to control after such a transplant procedure. It has also been speculated that the response to transplantation may depend upon the microbial and genetic composition of both recipient and donor prior to the procedure.[38] As yet, sadly we do not have a magic pill to reconstitute our microbiota and make us a healthy weight.

Epigenetics and obesity

By now, it should be apparent that there are many factors present from birth that influence our appetite and our weight, like the genes we are born with, the microbes we acquire as we enter

---

* Other indirect actions include effects on other systems, such as endocannabinoids – endogenous substances chemically similar to cannabis compounds – that have been found to have a role in the regulation of food intake and energy expenditure.

the world, and who we surround ourselves with. But it may be that, even before birth, there are environmental factors too that profoundly change the core of our biology and ultimately define our weight as adults.

In the 1980s, medical researchers noted a curious finding. As countries grew more prosperous, rates of cardiovascular disease rose. However, this was not simply a matter of heart disease being a disease of the wealthy. In fact, quite the opposite. In those wealthy countries, it was the poorest who had the highest rates of disease. This observation led to a theory, subsequently termed the Barker hypothesis, that exposure to an adverse environment while still in the womb, followed by an abundance of food in adulthood, might constitute a powerful contributor to certain diseases in later life. Supportive evidence for this view – that our uterine environment might have long-lasting effects on our health decades later – came from an experiment of human nature.[39]

In the summer of 1944, Allied troops marched across western Europe, rapidly recapturing much of France, Luxembourg and Belgium from the Nazis. As they entered the Netherlands on 14 September, the Dutch population expected liberation to be only a few days away, having seen how quickly their western neighbours had been freed from occupation. In support of the Allied offensive, the Dutch government-in-exile called for a national railway strike, but this call to action inadvertently resulted in immense suffering.

The rapid reconquest by the Allies unfortunately came to a sharp halt when they failed to gain control of a bridge at Arnhem, the crossing point for the Rhine river, blocking their progress northwards. The Germans banned all food transports into the country in retribution for the call to strike, resulting in massive food shortages for Dutch civilians. Despite the ban of food transports being lifted in early November, a harsh winter left most of the canals frozen, resulting in an inability to bring food by barge from the rural east of the country to the urban west. The west of the country, including Amsterdam, starved. Official daily adult rations fell from 1,800 calories in 1943 to below 1,000 calories per day in November 1944, dropping to a nadir of 400–800 calories between December 1944 and April 1945.

This severe rationing affected everyone, independent of social class, wealth, or indeed, pregnancy; children under the age of one were relatively spared, and while pregnant and breast-feeding women had entitlements to extra food, at the peak of the famine, these ceased. In the post-war period, those who had lived through the 'Dutch Hunger Winter' experienced growing prosperity and an abundance of food. Even by June 1945, daily rations had risen to over 2,000 calories.

Long-term monitoring of those babies born in the aftermath of the famine produced some striking results. Babies born to those mothers who starved during mid- to late-pregnancy had low birth weights and continued to be small into adulthood, with lower rates of obesity than those in utero before or after the famine. However, those babies born to women exposed to the famine during the early stages of pregnancy had healthy birth weights, yet went on to have significantly higher rates of obesity and cardiovascular disease in adulthood than those born before or after the famine, or indeed those exposed to maternal starvation later on in pregnancy.[40] These findings implied that the uterine environment at critical periods of development may result in permanent changes to our physiology, that prime us for good or ill health later in life. Insufficient nutrition during very early development may give rise to changes within our vital organs, on a cellular and metabolic level, that maximise our chances of survival after birth in a nutritionally deficient environment. But those very adaptations may actually have major detrimental effects on our health in a food-abundant world.

While early studies based upon the Dutch cohort focused on maternal undernutrition, it has also become clear that maternal overnutrition may also have substantial repercussions. Obesity in pregnant women has long been known to be associated with babies born large for gestational age. More recently, though there is increasing evidence that maternal obesity has a causal role in the development of obesity in the offspring. The higher the maternal pre-pregnancy BMI, the greater the degree of childhood obesity. The greater the degree of maternal weight gain over the pregnancy, similarly the higher the likelihood of childhood obesity.[41] These associations are also true when these children become adolescents and adults. As with the Dutch famine study, the timing of maternal weight change within the pregnancy has differential effects. Weight gain in early rather than late pregnancy seems more important for the precipitation of adult obesity in the unborn child.[42]

Of course, there are many potential confounders in these kinds of studies. In light of what we know about the role of genetics in obesity, the child may have inherited obesity-related genes from the mother. Or dietary and physical activity norms within a household may predispose both mother and child to obesity. However, animal studies refute the view that all this can be explained by obesity genes or household eating patterns. These experiments appear to confirm that something happens to the unborn child in the womb in response to the maternal nutritional status, that persists throughout that child's life. In rodents, inducing obesity in mothers through overfeeding causes obesity and other changes like increased blood pressure and eating-related behavioural changes in the offspring,[43] and indeed there have been similar studies performed in primates. In these animal studies carried out in the controlled environment of a laboratory, genetics, physical activity and dietary habits cannot be the explanation.

If this is truly the case, that maternal obesity or starvation in pregnancy drives obesity in children, how does this happen? As we have already heard with respect to wrath, one way in which our childhood experiences might give rise to personality disorders is through epigenetics – the modulation of our genetic code by our environment.

These animal studies point to these epigenetic mechanisms occurring in the children of obese mothers, including changes to genes that influence the regulation of appetite and energy balance by the hypothalamus. Diet-induced obesity in rodent mothers appears to permanently alter the response of the hypothalamus to leptin and long-term regulation of appetite in the offspring.[44] Epigenetic changes also appear to alter the activity of the leptin receptor gene and other genes that regulate satiety.[45] In short, in the controlled environment of a laboratory, devoid of influences of potential confounders like obesity genes or eating norms, the consequences of maternal overfeeding clearly affect the functioning of this system so crucial to the regulation of appetite and body weight, through epigenetic mechanisms.

Even beyond epigenetics and its effects on the leptin system, maternal obesity may impact the how the hypothalamus controls appetite in other ways too. It may programme the hypothalamus of the offspring through causing other developmental changes – predisposing to inflammation, altering the number and type of cells, even triggering dysfunction of mitochondria (the energy-producing organelles of cells) in this region of the brain.[46]

Thus, the environment in which we develop, particularly in the early foetal stage, appears to crucially affect how the brain machinery that regulates appetite grows and functions, for the rest of our lives.

But what does it mean for our own children, or even grand-children? Could these epigenetic changes be passed on, affecting the appetite control of our descendants too? In theory, epigenetic changes, not directly influencing DNA sequence, should not be passed down the generations; in practice, though, some alterations can be carried through to offspring, leading some scientists to propose that these epigenetic changes may be contributing to the current obesity pandemic. Obese mothers beget obese children, who beget obese children themselves, all through epigenetics – so the theory goes. While transgenerational epigenetic inheritance clearly occurs in plants and fungi, there remains an ongoing debate as to whether it occurs in humans.[47]

The reward of food

As James, the man rescued from his shower cubicle, illustrates, to consider our appetite to be defined only by our satiety thermostat (how our brains define when we have eaten enough – whether because of genetics, anatomy or epigenetics) is clearly not entirely accurate. While some of these factors might be in the mix for him, the overriding impression is of someone whose relationship with food has been tainted by his mental health, a person who is eating for comfort in the fog of sadness.

We have all experienced eating that piece of chocolate, a slice of cake or some other treat, despite feeling full. While circuitry with the hypothalamus, and leptin at its core, regulates our food intake with energy expenditure to maintain a stable body weight over time, there is more to eating than just satiety. Food is not just about fuel, as any gastronome will vouch for. It is also about pleasure, reward, motivation.

From an evolutionary perspective, for foods rich in sugars and fats to act as potent rewards, to promote eating even in the absence of an energy deficit, confers a distinct advantage. It ensures that calorie-rich food is eaten when available, permitting energy storage in anticipation of periods of starvation. It also acts as a trigger for learning. For example, if we quickly associate a particular stimulus, such as the sight of a bee's nest, with the reward of sweet honey, this also confers an advantage to survival. Perhaps I should say conferred, since this evolutionary advantage has become a liability in the current age.

A second neural circuit is involved when it comes to eating, not simply the hypothalamus monitoring out levels of fat storage. The circuitry of pleasure and reward.[48] And just as changes in the hypothalamus can lead to overeating and obesity, so can alterations in this reward system.

The 'dopamine hit' is much written about in the context of social media likes, sugar or exercise. That rush of gratification with a bar of chocolate, the headiness of the first cocktail of the night. Dopamine's major role is to orchestrate processes of learning and memory to develop behaviours necessary for survival. The reward of pleasure is the fuel for those aspects of our lives that are fundamental to it – food, drink, sex – and dopamine is the mediator of motivation to seek these experiences out.

Several neural pathways that release dopamine all originate within the mid-brain – at the top of the brainstem and adjacent to the hypothalamus – and different circuits mediate different functions, not only pleasure.[*49]

From the perspective of reward, however, one of these pathways is of most relevance. This pathway – the mesolimbic pathway – originates from an area in the middle of the brain,[†] and projects very widely (Figure 5). From here, these neurones

---

* From the world of Parkinson's disease, we know that dopamine projects from the substantia nigra to the basal ganglia higher up in the brain, important for regulating motor function and learning skills. Another pathway projects to the pre-frontal cortex – the surface of the frontal lobes closest to the face, fundamental to decision-making and planning. Dopamine deficiency in this circuitry reduces reactions to external stimuli (as in schizophrenia), or causes a deficit of attention, as in attention deficit hyperactivity disorder.

† The ventral tegmental area.

cast towards regions of the limbic system, highly involved in the formation and processing of emotions. They also project to the hippocampus in the deep temporal lobe, associated with learning and formation of memory. And, most importantly, they project to the pre-frontal cortex and a tiny cluster of brain tissue at the top of the hypothalamus – the nucleus accumbens. It is these latter projections that mediate excitement, gratification, and the search for stimuli that reward us with pleasure. This system is the seat of learning to seek out hedonistic experiences, that from an evolutionary basis allow us to survive and pass on our genes. The nucleus accumbens is considered the reward centre of the brain.

Any stimulus – an event, an activity or an object – that provides us pleasure or happiness, crucially provides reinforcement to make us learn to seek out these experiences again. Those positive emotions associated with these stimuli are the key drivers of trying to repeat the experience, sometimes at high cost. Increase dopamine levels and this will heighten the pleasure associated with a positive reward; decrease dopamine levels and this results in anhedonia, a diminishment of pleasure.[50] Expose yourself to environmental cues that predict pleasure, and the dopamine-producing neurones will show bursts of activity; predictors of punishment suppress activity in these same neurones.[51]

Eating tasty food results in release of dopamine within this circuit,* in direct proportion to how much pleasure the subject reports from eating the food.[52] This release of dopamine does not simply relate to pleasure alone, however. In fact, on first exposure to a particular food, these dopamine-producing neurones fire with abandon. With repeated exposure to that particular food, that dopamine hit lessens, and slowly shifts so

---

* Within a region of this network called the dorsal striatum, closely linked to the nucleus accumbens.

that it is the stimuli associated with that particular food that give rise to a dopamine spike.[53] It brings to mind those viral social media videos of infants tasting chocolate for the first time, the smile of pure joy spreading across their faces, as dopamine floods their nucleus accumbens. With repeated exposure, they learn the association between chocolate wrapper, the act of opening the packet, and the dopamine spike that occurs in anticipation of a mouthful of chocolate. Essentially, over time it is the prediction of reward about to present itself that is associated with dopamine release, rather than the reward itself. Since the point of this pleasure system is to drive behaviours, to act as a trainer of actions conducive to survival, it makes sense that this system links closely to emotional and rational regions of the brain, such as the limbic system and pre-frontal cortex, respectively.

It is not as simple as dopamine driving the liking of food, however. The signalling of the anticipation of reward is not the same as the reward itself. This constitutes the difference between 'liking' and 'wanting'. This sounds very odd, as generally we want what we like, and like what we want. In certain settings this does not always appear to be the case though. The prolonged abuse of drugs can result in an intense wanting of that drug, without the enjoyment of it once it is taken. Many drug addicts describe their addiction as a wanting rather than a liking. They need the hit, searching it out at all costs, rather than enjoying it.

Chocolate is addictive

Dopamine is not the only chemical in the brain involved in reward and pleasure when it comes to food. Particularly when it comes to the 'liking' of food rather than the 'wanting', other neurotransmitters also play an important role.* For example, the release of endogenous opioids (naturally produced chemicals within the brain simulated by heroin, morphine and fentanyl) in the amygdala conveys the emotional aspects of food. Curiously, in both humans and animals, sugar is an analgesic. Eating sugar lessens pain, suggesting that sugar may directly cause the release of endogenous opioids both in reward circuits but also those areas of the brain modulating pain.[54]

If that is indeed the case, then perhaps dieting may induce a mild withdrawal syndrome, potentially explaining why it is so difficult. If your brain is used to regular hits of morphine-like substances released by eating chocolate, when that stops happening due to dieting, you might crave that missing hit. Withdrawing from sugar bears a passing resemblance to heroin withdrawal.

Naloxone is the antidote to opioids, given by ambulance crews and emergency departments around the world to reverse the effects of heroin or fentanyl overdoses. In rats given sugar-rich diets, administering naloxone induces a withdrawal syndrome that is not seen when the same drug is given to rats fed a normal diet.[55] This implies that sugar, and possibly other foodstuffs rich in caloric value, may induce a type of addiction. It may also in part explain the phenomenon of emotional eating, the tendency to overeat in response to anxiety, irritability or depression, which is seen in about 60 per cent of people with obesity.[56] The release of these neurotransmitters in response to these sugary or fatty foods may be an effective form of self-treatment for low mood or anxiety, making people feel happier and calmer, albeit with very problematic consequences.

* * *

---

* These include neurotransmitters such as opioids, cannabinoids and GABA, but also hormones secreted by the gut.

Given the growing evidence of these dopamine-producing neurones, and the production of other chemicals like opioids, mediating the wanting and liking of food, could it be then that this neuro-machinery may go awry in some people with obesity – like James, whose relationship with food was so obviously unhealthy? Using a form of brain imaging called PET, where a radioactively labelled substance that binds to dopamine receptors is injected into the bloodstream, obese individuals can be shown to have a reduced ability to detect dopamine in their brains, with that decrease being proportional to their BMI.[57] This suggests that those individuals may require a bigger hit of dopamine to get the same effect, since their brains are less sensitive to dopamine.

A series of studies have demonstrated that these circuits involved in reward – the limbic system and other areas of cerebral cortex – show a greater response to food (or even pictures of food) in obese and healthy-weight individuals compared to lean people.[58] However, in obese individuals, while many regions involved in this circuitry show increased activity in anticipation of eating, they experience less brain activity than healthy-weight individuals when they actually eat food. There is a mismatch between the anticipated reward of eating and the actual consumption of food, which might be expected to drive compulsive eating. If you anticipate that a piece of cake might give you pleasure, and it does not quite hit the spot, you are likely to eat another slice. Again, there are parallels with drug addiction, where the anticipation of a drug is not matched by the enjoyment of taking it.

Indeed, this may be of relevance not just for James, but also for Alex and her PWS too. Those areas seen to be more active in the brains of people with PWS when eating include the nucleus accumbens, that 'reward centre'. This suggests that increased appetite in PWS may also be associated with an increased reward from eating or the anticipation of it, not purely hunger. In keeping with drug addicts, with their increased 'wanting' but not 'liking', Alex does not articulate a particular love of food. She seems to want it, exhibiting an indiscriminate need for it, rather than enjoying it. This chimes with Kate's comment describing Alex's relationship with food as being 'like a drug addict'.

We have already seen that maternal diet during pregnancy may influence the appetite thermostat. But there is also preliminary work to suggest that a mother's diet in pregnancy may also influence this 'hedonic thermostat' – the degree of pleasure or reward obtained – when it comes to food. Some researchers have hypothesised that maternal diet in pregnancy may also predispose to eating disorders in adulthood through changes to the reward value of food, driving an increased perception of sweet taste, increased pleasure derived from high-calorie foods, or other mechanisms.[59]

Why humans and grizzly bears are not the same

As with many areas in neuroscience, there remains the question of which is the chicken and which is the egg. Obese individuals show increased anticipation of the reward of eating, of the neurobiological drivers of wanting food. There are two possibilities for these findings, however.

The first is that altered brain circuitry drives appetite and obesity. These changes to the leptin syndrome, the functions of the hypothalamus and the limbic system, are the underlying cause of gluttony, of weight gain and obesity. As we have seen, there is a massive array of evidence to point towards this. In both healthy individuals, and those patients and animals with damage to these brain systems, there is incontrovertible proof that the brain controls our weight.

There is an alternative explanation, though. Perhaps obesity causes these changes to our brains too. Perhaps some of these brain changes seen in the obese are secondary to massive weight gain. Do these differences in the reward system therefore cause obesity through predisposing to abnormal eating patterns, or are some of these changes subsequent to obesity, resulting from abnormal eating habits?

Even when it comes to the leptin system, which is relatively well-characterised, this is not entirely obvious. Like the grizzly bear in the autumn, obese individuals have higher levels of leptin than normal, since their excess body fat is producing large quantities of it. Therefore, this system should be actively suppressing appetite, signalling that the stores of energy are more than adequate. Instead, the effects of leptin are muted. Even giving obese people analogues of leptin does not suppress their appetite[60] – again implying a relative insensitivity to leptin.

But unlike the grizzly bear just before hibernation, in humans this leptin resistance does not suddenly reverse with the change in seasons. It continues unabated.

How this insusceptibility to leptin in humans, and its failure to suppress hunger, develops remains a mystery. It may of course be that a relative insensitivity to leptin is what has fuelled their appetite in the first place, and led them to be obese. But another possibility is that obesity in and of itself may contribute. It has been suggested that obesity might directly block leptin from penetrating the brain from the bloodstream. If that indeed is the case, then it worsens the position of humanity in its current state of an obesity epidemic. It implies that being obese may make further weight gain inevitable, that when you get to a particular weight you are on an almost irreversible path to worsening obesity. The result is a potent vicious cycle of brain dysfunction causing weight gain causing brain dysfunction.

* * *

As Alex's parents and I draw to the end of our discussion, I ask if they have more general reflections of what life caring for a child with Prader-Willi has been like. They are keen to voice something that, at the current time, with the NHS crumbling under the weight of demand, is too rarely heard. 'It really helps us to appreciate the NHS, the fact that the NHS is paying for every single referral or specialist. I think it makes the UK a real beacon, relative to somewhere like the US, for people with complex syndromes.' They say that the NHS constitutes a safety net. At the back of their minds, they are reassured that, should anything happen to them, Alex will be taken care of in a way that is more difficult to achieve in many other countries. But they also talk of the distress of being told about the condition in the way that they were, leaving them initially bereft of hope. Kate now talks to medical students yearly about breaking such news, and has also set up a group for parents with children with special educational needs and disabilities that meets regularly as a forum for support and passing on of knowledge. Jon and Kate recognise that, in the world of the internet, and with charities such as the Prader-Willi Syndrome Association (PWSA), information is easier to come by than it was when they first learnt of the condition. Since that time, Alex has graced the cover of a PWSA information pamphlet.

And there are reasons to be hopeful for people diagnosed with Prader-Willi syndrome. In the past, they would often die young, largely due to the complications of obesity. Alex has been on regular growth hormone injections from an early age, and a number of new drugs like the GLP-1 agonists, those treatments for diabetes increasingly being used for weight control, are now available or going through clinical trials.[61] As Kate points out, trying to understand how satiety signals malfunction in people with PWS will potentially be of great benefit for wider society, and the growing epidemic of obesity.

Clarifying the extremes of the human experience aids our understanding of the broader range of our nature.

I received some very sad news while writing this book, however. Despite the optimism for better treatments for people born with PWS in the future, for Alex this will come too late. Shortly after she and I met, in the summer of 2023, she unexpectedly passed away. She was surrounded by her family. Her cause of death was found to be a spontaneous rupture of her stomach. While this is described in people with PWS, it is usually caused by overdistension due to overconsumption of food. This was not the case for Alex.

* * *

The stigmatisation of obesity is widespread, deemed a failing of character, a lack of moral fibre, a physical or mental laziness. Of course, for most of us, obesity, appetite and eating behaviour cannot be reduced to one or two neural or hormonal circuits alone. It is a complex, multifactorial process, with contributors from genetics, environment, psychology and societal factors. I see it in those individuals who undergo obesity surgery, rapidly shifting huge amounts of weight before often gradually gaining it again over the subsequent years. Addressing only one facet of this cocktail of factors is often destined to fail. While most people undergoing this type of surgery are evaluated by nutritionists, physiotherapists, psychologists, psychiatrists, as well as the surgeon wielding the scalpel, it is surprisingly challenging to maintain a long-term healthy weight.

But there are distinct explanations for why people living in the same society, surrounded by the same access to food, burning the same number of calories a day, may have vastly different appetites and weights. Those evolutionary imperatives – to burn energy efficiently, to seek out high-calorie foods even when sated, or possibly to keep us warm in colder climates – are now conspiring against all of us in the modern world. It is just that this modern world is conspiring more vigorously against some people than others. Losing weight is not quite as simple as just reducing what you eat through willpower.

And the truth is that 'gluttony' is not a moral act, a failing of our 'souls'. Our appetite is a consequence of our genes, our guts and those functions in our brains that control our hunger and mediate the reward of food.

3.

Lust

But virtue, as it never will be moved,
Though lewdness court it in a shape of heaven,
So lust, though to a radiant angel linked,
Will sate itself in a celestial bed
And prey on garbage.
William Shakespeare, *Hamlet*, Act I, sc. v

'Since my son's discharge from the Army . . . his language is obscene, and his chief conversation is about sex; this began from his first conscious moments in hospital when he said all the nurses were prostitutes.'[1] Thus a father described his son, a young nineteen-year-old private in the British Army, wounded on 6 August 1944. In the midst of battle, a bullet, visible on the X-ray of his skull, had passed through the front of his head, leaving a large wound on his right forehead. The foreign body, some 3 cm long, had lodged deep in the right frontal lobe of the brain.

On admission to hospital, three days after the injury, he was paralysed down the left side of his body, and was sleepy and monosyllabic. Remarkably, within a few days, he was fully conscious, knew where he was, and could accurately describe his home, his regiment and career in the army, albeit a little confused about recent events. An operation to remove the bullet on 21 August was successful, and although he was occasionally incontinent of urine, he was discharged some three months later, having made 'a good physical recovery'.

While certainly physically recovered, mentally it was very much otherwise. By 1947, he was on probation for several charges of burglary, and his father reported a marked change in his personality. 'He is unable to remain still for any length of time . . . He sleeps abnormally long hours and has an enormous appetite . . . Notwithstanding the fact that sex predominates his mind, he has not shown any zeal or over-interest in running after girls, and I have no reason to think that I should have any worry about this.'

Hugh Jarvie, a neurologist at the Radcliffe Infirmary and the Military Hospital for Head Injuries, Oxford, described his first meeting with the young man in 1951. After a few minutes of silence, the patient launched spontaneously into 'sex talk', opening the conversation with: 'You see, it's the sex problem.' He began to describe encounters with 'a woman of undesirable character', and seemed obsessed with the sexual proclivities of American servicemen.

Prior to his head injury, Jarvie noted, the soldier could hardly have been considered a 'man of the world': his first sexual experience had happened at the age of fifteen, when 'a young girl allowed him to handle her breasts'. He was wracked with fear that she would tell her parents. His terror was only compounded when she told him she was menstruating, illustrating his lack of understanding of basic biology. At seventeen he had been 'induced' into sexual intercourse by the wife of a serviceman who had been sent overseas. Apart from that, his only other sexual activity had been limited to, in the language of 1950s' Britain: 'Cheap American magazines of a type describing sex play.'

His head injury had had a lasting impact on his behaviour.

His speech was filled with uninhibited conversations about sex, torrential and unprompted. Alongside his fixation that all the nurses were, in fact, prostitutes, streams of 'sex talk' constantly emanated from his mouth: 'Yanks are beasts,' he said. 'Every girl . . . is carrying a child by a Yank. Two of them raped a girl of fifteen. Sex is my dominating feature. I like sex books – *Lady Chatterley's Lover* . . . I wonder if Lady Chatterley ever lived, the most filthy-minded bitch. I read Hank Janson [a pulp fiction writer of the era]. That's a new one on you. Always girls. In his books, men are always mixed up with some woman's blouse or thighs . . . I know I am a nuisance. I can't keep off sex talk.'

Yet despite his invective, it was clear that though the young man was altered in the content of his speech, his sexual behaviour was unaffected. While there had been sexual encounters before and after the injury, there had been no physical loss of control. Rather, it was his ability to judge and conceal his thoughts during normal interactions that had been left impaired. 'In spite of his undiluted sex talk,' his father explained, 'he is really rather prudish. Anything on a film upsets him and he will come out if he feels it is slightly indecent. He was always rather a modest type of boy. He would not enter a bathroom where his sister might be washing or where she might be slightly unclothed.'

Therefore, while he was totally uninhibited when it came to talk, this was not followed through by action. In fact, Jarvie surmised that most of the young man's topics were actually protests against the physical act of sex – a sort of moralistic rally against the 'bestial nature' of men and women. He proposed that the brain injury had simply revealed his long-existing preoccupation with sex, as illustrated by his interest in pornographic magazines prior to the war. What had been previously suppressed, concealed behind a veil of social norms, could no longer remain hidden in the context of his brain damage. In essence, the brakes that had prevented his long-standing views on sex bubbling to the surface had been obliterated by that bullet.

While in this young private's case it was the expression of thoughts of sex that was influenced by his brain injury, other cases cited by Jarvie in the same paper presented examples more dramatic in nature. Another soldier, a twenty-four-year-old, had been injured at around the same time, a piece of shrapnel ploughing through his skull, dragging bone chips in its wake, destroying his frontal lobes and finding its final resting place behind his right eye. The eye had to be removed along with a suctioning out of the cavities in his brain. On admission to hospital in Oxford, he was drowsy and irritable, and had no memory of going to France or of his injury. His recovery was stormy, with infections, a seizure and further operations, but eventually he was discharged, some six months after being wounded.

It was only in 1951 that this former soldier came back to medical attention, largely as a result of his wife's distress. She described a marked change in his behaviour. Prior to the war, she said, he had been rather reserved, a man of a quiet disposition. He had grown up in a home with a dominant mother, bullied at school. Sex as a subject of conversation in the family had been taboo, never discussed. Now, however, he would beat her and the children, would pilfer inconsequential items at work and had become rude and generally difficult.

By far the most unsettling aspect for his wife, though, was not his violence or stealing. 'Perhaps the most trying part has been his attitude towards sex. He is not content with having intercourse once,' she reported, 'but will try to go on all night, and has said on occasions that he would like to treat me as a woman of the streets and rape me.' The medical examiner noted: 'The wife stated that he was always sexually demanding even before injury but that his demands have now intensified. Instead of coming home from work he goes to the local public houses and discusses her and his relations with her in public. He masturbates openly in front of her and keeps saying he should have a love affair with someone of seventeen.'

The man made no attempt to deny his wife's claims, simply stating that since his brain injury it had been more difficult to control himself, both his sexual behaviour and his temper. Despite the devastating nature of his brain damage, his intellect, however, was completely unaffected. As with the earlier case, there were hints that in a previously overly inhibited individual, with a tendency to be 'sexually demanding', the damage to his frontal lobes had unleashed what had previously been stifled.

In a third case from Jarvie's series, he described another veteran – this time a twenty-five-year-old sergeant, injured at the tail end of the war, in February 1945. A foreign body had pierced his left eye, before one fragment had crossed his brain to its final resting place in the right parietal lobe, towards the upper rear of the brain. The damage left him with weakness down the left side of his body, drowsy and partially conscious, but he made a reasonable recovery before being discharged from the hospital a few months later. However, within a matter of months, he was up in court on two separate counts of indecent exposure, and subsequently served several custodial sentences for similar offences in various parts of the country.

The man denied any 'exhibitionism' before his injury and, as with the prior case, told doctors that he had been brought up in a strict home where sex was a taboo subject. He described himself as 'shy of girls'. But since his discharge, he recognised that he had lost control over his sexual behaviour, and indeed Jarvie reported him blushing as he related the incidents that had led to his convictions. His father was recorded as saying: 'He is not the same man at all as he used to be before his injury, being then kind and obliging at all times. He was a boy who never gave his mother or myself any worry at any time. He had a kindly disposition and was as honest as the day. There is no doubt about it, he is a changed person since his knock on the head.'

The sergeant described how his offences usually occurred after heavy drinking, and that when he exposed himself: 'He was overcome with strong sex feeling which required relief. He stated that his sex feeling had increased since the wound; his wife told him that "sex was all you married me for" and complained of the roughness of his sexual approach, although she had made no such complaint in the early months of the marriage, which had taken place in 1943.' He said of himself: 'As a young man I was not overkeen on girls and was always uneasy in their presence. Now my wife refuses to live or sleep with me and complains that I persistently wake her up during the night to have sexual intercourse.'

In all three of these soldiers' cases, damage to their brains had somehow impaired their ability to control their sexual impulses, either in word or deed – demonstrating the absence of neurological mechanisms for restraining 'lust'.

* * *

Jarvie's patients are far from the only individuals to exhibit changes in sexual behaviour in the setting of neurological damage or disease.

The first thing that strikes me about Simon* aside from his stature (he is well over 6 feet tall) is his voice. He speaks very softly, not exactly in monotone, but broken by long pauses, as if he is processing the impact and significance of each and

---

* Name changed.

every word. I cannot initially fathom whether he is simply shy, whether he is riven by shame or if there may be another reason.

A thoughtful and likeable young man of only thirty-four, he is, despite his hesitancy, brutally honest. His flat is a short distance from the River Thames. In one room I can see his home office, a computer and IT equipment – evidence of his job as a software developer. As I step into his living room, I notice a play zone scattered with toys – clearly his young daughter's territory. In spite of Simon's youth, though, he has already been diagnosed with Parkinson's disease – rare at this tender age – and I wonder if that might explain his slow and deliberate speech. In other respects, however, I see no mark of the condition upon him. He has no tremor, and he moves and walks fluidly, without stiffness or slowness of movement – the usual hallmarks of this disorder.

'I started getting symptoms at the beginning of [Covid] lockdown,' he tells me. 'When walking along by the river, I noticed that my left toes were curling.' Over time, his foot worsened, and it became more difficult to walk. He began to notice other symptoms, too. 'My left side felt a bit floppy when I walked. I went to a physiotherapist, who said it was probably just a pinched nerve or something.' He began tripping over his curling toes, his left shoulder rose up as he walked, and his left arm and leg felt just a little slower, a little stiffer.

It took him about a year until he went to see a neurologist, by which time he had developed a subtle tremor on his left side. A tentative diagnosis of a more benign form of tremor was made, but with the passage of time, a movement disorders specialist has more recently made a formal diagnosis of Parkinson's disease.

His daughter would have been about two years old then, and I ask him how he took being given the diagnosis. 'When I was researching these symptoms, I saw a lot of similarities with Parkinson's disease. Some people mentioned their toes curling and lack of arm swing. But everything I gathered online was just depressing. I was trying to avoid researching it too much.'[2]

For Simon, receiving the diagnosis of Parkinson's in his early thirties was not as devastating as I would have expected. He had been worried about something even graver. Around the time of the onset of his symptoms, and the birth of his daughter, he had become increasingly preoccupied about his health, developing a severe health anxiety. His doctor had prescribed him medication to bring his anxiety under control. The drug had eased the obsessional thoughts about his health, but as his physical symptoms had worsened, he had decided to come off his medication. 'I started to get paranoid that it was something worse. One of my big fears is ALS [motor neurone disease]. I started to get convinced I had that. Maybe it was because I had stopped the sertraline [an antidepressant and anti-anxiety medication] a little before then, but I started to get obsessed with checking my legs and my strength. It was taking over my mind. So it was almost a kind of reassurance that it was Parkinson's.' He hoped for medication that would manage his symptoms, if not cure his disease.

Parkinson's disease is not the primary reason that I am here, talking to Simon, however, although his condition may be related to it. The main focus of discussion is his change in behaviour. It is the reason for his partner leaving him, his friends ostracising him, for his isolation and the discomfort I detect in his voice.

Simon has not strayed far from his origins. He was born in West London, close to where he currently lives. His love of computers was evident as a young age. 'We lived in a big house, and had tenants on the floor above us,' he says. 'One of them had a laptop with computer games, and I became obsessed with going up [to the flat above] and playing them.' His father bought him his first computer at the age of ten, in an effort to avoid his young son irritating the tenants. 'So then I just become obsessed with it. But there weren't that many games, and they were expensive. I found more fun in playing with all the settings on the computer and finding out how to do everything on it.' He rapidly became an expert and recalls that he would be teaching his peers in computer lessons at school.

He quickly realised that academic studies were not for him, and that his passions lay in digital art. After an initial exploration of the artistic route, he went back to college to learn graphic design and computer graphics. A further course in 3D graphics resulted in him getting a job at a computer games company, but after a year, the recession brought his career to an abrupt halt. After a couple of years back at his parents' – 'just playing games again – my parents must have thought I was hopeless at that point' – he managed to get another job, then another, and ultimately ended up coding for a financial software company in the City of London.

Throughout the ups and downs of his schooling and employment, Simon has been in a series of long-term relationships. 'I lost my virginity at the age of fifteen . . . I think I next had sex with my long-term girlfriend at the age of sixteen.' He was together with this young woman for eight years, stable, faithful and unremarkable. Their sex life was unremarkable too. Active, but not atypical in any way. 'I did feel like I wanted it a bit more, but since it was my first girlfriend, I didn't really think any more about that. Good and stable. It was pretty vanilla. I probably would have been open to trying other things, but we didn't do anything.'

I ask what happened to terminate that relationship, and he describes meeting someone while on holiday: a woman from Eastern Europe, with whom he became, in his words, 'rather fascinated'. He ended his relationship, and embarked on a long-distance romance, culminating in her coming to the UK to live with him. Again, a healthy romantic and sexual relationship. But pressures of work eventually took their toll. 'She was studying during the day and working in catering at night. I was commuting to my first job in software, driving an hour and half each way. We barely saw each other. Our friendship circles changed. She developed her own, and I stuck with mine, and we just grew apart.'

A shift in his relationship with sex came with his next liaison. He remembers always initiating sex with that partner and wanting to be with someone who wanted to initiate sex with him. 'I began to obsessively research online how to have sex when people don't want to have sex with them [i.e. how to make himself more sexually alluring to his partner],' he adds nervously. 'I started to think more and more about sex, to fantasise about having sex with someone else, and had a compulsion to look up sex stuff online.'

I ask him if he actually did want to have sex, or whether it was simply that thoughts of sex would constantly enter his mind. He answers tentatively and uncertainly. 'I just felt a kind of compulsion. If she would leave the house, I would go online. Almost like . . . instinct. I felt like it was just building up, something that I needed to do.' Sometimes for the whole day, if he had the opportunity. The use of porn would quieten his mind, lessen his ruminations for a short period of time, before obsessive thoughts of sex would return. 'I was quite ashamed of it. I didn't tell anyone about it. I just felt that I was being quite disrespectful to my partner, and felt bad about it.' He makes an almost throwaway comment: 'It was at the same time as I had this similar compulsion to google health issues. It coincided with my health anxiety rising. What I found interesting was that exercise, when I got into it every now and then, would dampen my sexual obsession and my health anxiety. It kind of balanced those things out.'

I ask if he felt he could control these thoughts, these 'urges'. His answer is rather unexpected. 'I don't know. I think if I could have stopped it, I probably wouldn't have taken it further when I was in London. I started to go to those massage places where they would . . . basically . . . masturbate you.'

This change in behaviour was not sudden, however; it developed gradually. He says that he started doing this a few years into the relationship, after watching porn on a regular basis. He began to be obsessed with the idea of massage parlours, having heard that they would offer sexual services. 'That just put a seed in my head. I just felt like I needed to do that. A really strange compulsion to do it. I guess it was something I thought about loads before actually doing it. Initially, I would be like: "No, why would I do that? That would be embarrassing. How bizarre is that?" Then one day I did it. And I felt terrible afterwards. But I also felt like all the anxiety, the worrying leading up to it . . . it was almost like a release afterwards. On the one hand terrible, on the other much happier. It was very confusing.'

Simon uses the terms 'compulsions' and 'obsessions' throughout our discussion. He later says of his £20 trips to the massage parlour near his work: 'It was a horrible experience going into it. But for some reason, I felt it was a track I was on, a ledge. An obsession that I had to get over, otherwise I wouldn't be able to stop thinking about it. It was something I had to get rid of. There was initial relief – "Now I don't have to worry about it, it has gone from my mind." But then it would just slowly build up in my head again,' he says. 'And I'd get this urge to do it again, and again. If I didn't, the obsession would just be stuck in my head. I wouldn't be able to shake it.'

His ruminations on sex continued. Thoughts of having sex with strangers endlessly circulating within his mind. He describes an increasing build-up of stress, of obsession, like the magma chamber of a volcano boiling under the surface. He would scroll through personal adverts for sexual services on his phone. 'It seemed kind of fantastical, but it quickly became an obsession. I would build up the idea of that happening.' Soon, though, these fantasies became realised. More often than not, his actions would be limited to thoughts and the perusal of adverts, but the introduction of alcohol sometimes made these ruminations more intense. It is clear that alcohol was a catalyst, the stoking of his obsessional fire. 'The thoughts were always there, but it was something made worse by going out and having a drink in Central London. It just made all these emotions spike.'

After a few drinks on his way home from work, Simon would pray for intervention, divine or otherwise. 'I would hope that a reason would come up to stop me from doing it. Perhaps my train coming quicker, or someone to call me. I knew it was wrong. I couldn't stop myself. It was such a strange feeling.' Without reason not to, Simon would occasionally follow through on his thoughts, visiting sex workers in their apartments, and paying for sex. Again, the realisation of his thoughts left him with clashing emotions. There would be the release of tension, the stilling of his mental machinations, a feeling of clarity. But also, feelings of shame, of guilt and the ridiculousness of his actions. 'I wanted to laugh at myself, for being so stupid, for being so obsessed about it.'

I ask whether he experienced sexual gratification, a pleasure in the sexual act. He is very clear: 'It wasn't great. It didn't feel like much. It wasn't particularly enjoyable. It felt good because the build-up of those thoughts, the constant thinking about it, I got rid of that, which made it worth it in the end.' He says he was not particularly attracted to the women he paid for sex.

From his description, the ongoing mental churning was distressing, and the transient stilling of his mind obtained by paying for sex was the principal reward.

His ongoing relationship broke down, although he had kept this aspect of his life hidden. Simon met someone else and assumed that, now he was happier with his new partner, that would be the end of that kind of behaviour: 'That was the reason [for the trips to the massage parlour and sex workers]; it was that I was with a person who wasn't right [for me].' After a period of time in this new relationship however, his familiar thought processes, what he describes as his 'weird anxieties', resurfaced. He tells me that he only followed through on his sexual thoughts once in this new relationship though.

He and his partner got closer and closer, and they moved in together, just before the pandemic. 'So, everything like that [paying for sex] would have stopped anyway. And then she got pregnant with my daughter,' he tells me. 'After that, I just thought: "I'm not going to do that again," because it just felt different then.' Throughout this time, he remained on his anti-depressant, the drug initiated to treat his health anxiety.

About a year ago, however, he came off this drug. 'We were at the pub, with friends. One of the group had just been through a break-up. And we were just chatting, and she just asked me if I wanted to have sex with her. It fed on all my insecurities about people not wanting to have sex with me, and I was a little drunk. And I just said yes.' They slept together that night.

Simon ascribes this event in part to his coming off his medication a few weeks before. His emotions had been rather volatile. 'It was exciting and made me feel really good. But I felt my brain broke a bit after that, with the confusion of the day. I felt happier, but also very sad. Because of what it meant. I had taken a thousand steps back.'

He began to ruminate over his sexual betrayal, the guilt and shame overcoming him. For a week he fixated on telling his partner about his actions, mulling it over and over. The act of telling her, of confession, became another obsession. 'I wanted to be honest with her, because I had never told anyone about [these thoughts and deeds]. I just thought I need to deal with this. Otherwise, it will be going on for ever. I knew it would be bad for our relationship,' he says. 'I thought maybe there was a 10 per cent chance she would be okay with it, and she would understand. Yeah . . . that wasn't the case. She didn't take it very well, and then left.'

* * *

Why not sow one's seed with abandon?

Lust, not love. Unbridled sexual desire. Animalistic, primaeval; the irrepressible urge to engage in the act of procreation; to physically possess. In a biblical sense, the desire for anyone other than your spouse in the eyes of God. In the modern world, perhaps a more forgiving view of lust: a sexual desire heightened to the extent that it causes problems for yourself, or the people around you.

It is predictable that this 'sexual beast', this inner drive for sex, should reside in all of us. After all, sex is the engine that drives the survival of life and the Darwinian process of evolution. The etymology of 'genitals' is from the Latin pertaining to generation or birth. Without it, there is no reproduction, no passing on of genetic material, no intrasexual competition − where inherited qualities linked to success in evolutionary terms are more likely to be passed on to future generations due to competition among potential mates. Arguably, every fibre of our anatomy, physiology and behaviour is finely honed to complete this one act: of sex and reproduction. Without sex, life as we know it – and indeed life for all but asexually reproducing organisms – would cease to exist. That a drive for sex should be innate within us is to be expected.

Rather than asking why lust exists, a more pertinent question is why we humans should have developed machinery to limit our sexual instincts, those mechanisms that suppress our desire so obviously impaired in Simon, and Jarvie's patients. Surely the more sex we have, the more people we have it with, the greater our chances of reproduction?

Maybe indiscriminatory sexual intercourse, akin to what some of these brain-damaged soldiers aspired to, is not ideal, at least in a world without contraception. The right choice in whom to mate with brings with it multiple benefits: genes conferring survival advantages, such as a healthy immune system, provision of resources and physical protection. A poor choice has significant downsides: abandonment, sexually transmitted diseases, loss of reputation, a genetic inheritance that weakens your offspring. There are therefore potential explanations for a biological imperative to be selective rather than without discernment.

If that is the case then – that procreating at every possible opportunity may run counter to evolutionary principles – how do we humans, with normally functioning brains, unimpaired by shrapnel or bullet, actually select our sexual partners? What influences our choice of whom to sleep with?

In the real world, as Simon illustrates, it is readily apparent that any explanation of how we choose our long-term partner does not necessarily equate to who we have sex with.* In the

---

* Theories of mate selection abound in the historical literature. True to preconceptions of Freudian psychology, Sigmund Freud argued that people seek mates who resemble their opposite-sex parents.[3] Others proposed that the choice of mate came down to seeking qualities lacking in themselves,[4] or conversely, that people seek out mates similar to themselves, to ensure equitable exchange of resources.[5] A large 2023 study has found that, for long-term partners at least, for

domain of sexual theory, there is another thorny issue too – that this discordance between long-term partner and who we have sex with also extends to who we might procreate with. In clinical practice, we will often run genetic tests on people to determine the cause of their neurological disorder, even when the condition is known to have a clear pattern of inheritance. It is not unheard of for these genetic tests to produce rather unexpected results, that the father is not who it is thought to be. The most reliable studies suggest that non-paternity rates are somewhere between 0.5 and 4 per cent, although these do vary considerably between cultures.[7] Some older studies, dating back to the 1940s–80s, suggest much higher rates historically, up to 28 per cent.[8]

Any description of how we select our sexual partners, and the evolutionary drivers of these behaviours, therefore needs to account for the fact that, while frequently we choose a long-term partner to procreate with or have sex with, in human society, this is not always the case. Understanding the processes that underpin how we select our long-term mates, and those we have sex with, is crucial to demonstrating the function of those biological and psychological mechanisms that regulate our lust, and what is lost or compromised in the setting of neurological dysfunction.

---

a wide range of attitudes, educational attainment, substance use, psychological and anthropometric traits such as weight-to-hip ratio and BMI, opposites don't attract. Whether we seek out partners like ourselves, or whether we converge with time is unclear.[6] The problem with these theories is that they did not fully justify the purposes for why these patterns of mate selection might have evolved, what evolutionary advantage may be conferred by choosing a mate in the image of your father or mother, or someone similar to or complementing yourself. These older theories of mating practices also limited their discussion to long-term mating, such as marriage. They failed to account for the various contexts in which mating occurs. Not just marriage or long-term relationships, but also short-term mating, or even casual encounters.

Sexual strategies for men and women

In recent years, a new theory, termed sexual strategies theory, has been formulated to better explain the multifaceted nature of human mating psychology, and the search for evolutionary advantage.[9] Our mating behaviour has developed to maximise the chances of healthy offspring, to continue our lineage.

In our modern age, in the era of contraception, many of these evolutionary pressures that have driven our sexual behaviours have vanished. Nowadays, who we choose to have sex with seems far removed from those forces that determined whether we and our offspring live or die. But over the millennia and generations before, the direct links between sex and reproduction have moulded our genes and our sexual psychologies, throughout the evolution of *Homo sapiens* and even before it.

This theory proposes that both men and women have multiple different mating strategies, each attuned to the circumstances they find themselves in. Essentially, depending on our circumstances, we have the propensity to pursue long-term relationships, but also to chase infidelity and one-night stands. These different paths each have evolutionary underpinnings, with clear advantages in the battle to pass on our genes. Each mating strategy is fine-tuned to maximise the benefits and minimise the costs in each context. Furthermore, sexual strategies theory proposes that men and women will differ in their mating strategies, in part due to the obvious contrast between the time investment in producing a single child (nine months versus the brief act of copulation), but also due to differences in the challenges faced repeatedly during human evolution.[10]

The evolutionary case for short-term mating

That men should have evolved to seek brief couplings is perhaps to be expected, given the differences in the commitment necessary to procreating. Like Simon and his sexual encounters with the sex workers he does not feel particularly attracted to, where sexual gratification is viewed as 'low-cost', men will lower their requirements much more than women when it comes to a variety of qualities, including attractiveness, intelligence and personality. These differences in terms of attitudes to casual sex pervade every culture,[11] without exception.[12]

For women, the evolutionary drive for short-term mating is less immediately clear. The costs of reproduction, in terms of time and resources, is much higher. Nine months of pregnancy, followed by breast-feeding, nursing one's children – these are several orders of magnitude more of an investment than just sticking around for the moment of conception. For a woman, more sexual partners does not result in more children, unless the long-term partner is infertile. Yet it is clear, as these studies of non-paternity rates show, that female short-term mating is not a rarity. If we argue that our behaviour is fuelled by our evolutionary needs, it begs the question as to what has driven short-term mating strategies for women too.

When it comes to casual sex or short-term mating compared to a long-term mate, women place more importance on physical attractiveness,[13] and usually prefer more masculine male faces.[14] Women who have a tendency to recurrently have short-term partners – termed 'unrestricted sexuality' in the psychological literature – seem to emphasise physical attractiveness and sex appeal much more than other women.[15] This hints at a potential reason for this sort of mating strategy. Perhaps, these qualities that are sought in short-term mates represent a shopping around for good quality genes – those that confer fitness, in evolutionary terms: that sex appeal, physical attractiveness, is an indicator of genetic good health.

Evidence to support this 'good genes hypothesis' – that women seek out short-term partners to increase the genetic health of their offspring – is weak, though. This theory, of shopping around for good genes, implies that women might seek out genetically healthy partners for their offspring simultaneous with a long-term partner for support and resources. If the search for better genes is indeed the goal of women's short-term mating, then we would expect to see differences in preference for sexual partners at different stages of the menstrual cycle – for men with proxy markers of good genetic health at the time of the month when conception is most likely to occur. There is minimal evidence to support this, however. An analysis based upon the amalgam of fifty different studies has shown a very slight shift in preference for more masculine faces around ovulation,[16] but these sorts of studies are far from consistent, and many papers have shown no effect of ovulatory phase at all.

If there are doubts about the 'good genes' hypothesis, then what else might explain short-term mating in women? Infidelity or one-night stands for women may bring evolutionary advantages in other ways. One potential reason may be that these short-term relationships facilitate the 'trading up' to a better partner, testing the waters to see if there might be better mates out there, or even easing the escape from an existing relationship. Some psychologists have even proposed that short-term mating guarantees a back-up partner should anything befall the existing long-term partner – the so-called 'mate insurance function'.

Some findings support this view, that short-term mating in women is not necessarily to do with the search for better genes, but more to do with finding a better partner. Firstly, four-fifths of women report falling in love with an affair partner, compared to about one-third of men.[17] Since love is typically a feature of long-term relationships rather than one-night stands, it would be a redundant emotion if such encounters were primarily driven by the search for new genes only. Additionally, one of the strongest predictors of infidelity in women is relationship dissatisfaction, in stark contrast to men, and that factors which incline women to infidelity include meeting someone who is more successful than their existing partner, or someone who is keener on spending time with them, or having an existing partner who is unable to hold down a job. Indeed, in women who are unhappy in their relationships, they show more sexual interest in other men not only in the ovulatory stage of the menstrual cycle, but in other phases too. These facts all imply that infidelity is more about finding a new partner, a 'better' mate, rather than simply finding a new gene donor.

The evolutionary benefits of long-term mating

We humans are outliers in the animal world, however. While the human race clearly does sometimes seek out brief liaisons, generally it pursues another strategy. In contrast to other primates, human males sometimes dedicate enormous amounts of energy to being parents. Only about 3–5 per cent of mammalian species demonstrate long-term pairing akin to human long-term relationships.

In evolutionary terms, for women the benefits of a stable long-term relationship are more immediately obvious, in the context of the time and energy they have to dedicate to carrying and nurturing their offspring. Long-term mating potentially allows access to resources – women place more value on the financial prospects of their partners than do men, in a variety of cultures and religions.[18] Women are more than a thousand times more sensitive to salary when rating men than the other way round,[19] and even in hunter-gatherer societies, the ability to provide in terms of hunting and foraging is of huge importance to women selecting a mate.[20]

Another potential advantage is that of physical protection. Hence the consistent findings of women finding tallness, broader shoulders and muscularity more attractive. In a study of personal ads, a man's height was one of the strongest predictors for responses.[21] And these benefits, of a provider and personal bodyguard, extend to the woman's offspring as well, contributing to the likelihood that her genes will make it down to the generations thereafter.

But what about men? Surely the way to maximise the chances of passing on your genes to impregnate as many women as possible, to play the numbers game? What are the evolutionary imperatives for long-term relationships rather than the brief encounters that serve most of the mammalian world so well? Well, in part, it may be down to women. Ultimately, from an evolutionary perspective, a man also wants a mate of high genetic quality, to maximise his own offspring's chances of success, and to avoid investment in other men's offspring. The goal is to find a high-value mate, one that is genetically healthy and fertile, to monopolise her reproductive resources, and to ensure that the offspring that he thinks are his own actually are. And if women of higher genetic and reproductive value require higher commitment, then the strategy of choice is clearly to show that commitment. Essentially, the evolutionary drive for men to pursue long-term relationships is defined by the requirements of high-value women, some scientists suggest.

Concealed ovulation and reproductive value

As a male, the priority in evolutionary terms is to ensure one's offspring are many, and as healthy as can be. The strategy of sleeping with as many females as possible, getting many women pregnant, would certainly increase the number of offspring one has. But if a male cannot access high-genetic-value females using this shotgun approach, those with whom mating would produce the most genetically fit offspring, one must take a different tack. If those high-value females demand long-term commitment, a male may be better off pursuing this strategy.

Then how can males identify those women who have high reproductive value, those who have good genetic health and fertility?

Unlike many animals, and indeed our closest relative, the chimpanzee, ovulation in women is not readily apparent; female chimpanzees signal ovulation through a swelling and pinkening of their genitals. The lack of overt clues of fertility in human females creates two issues. The first is how to ensure that you, as a man, are not investing the energy and time in trying to have sex with women who are not ovulating. The simplest way is to have sex with the same woman repeatedly, to ensure that at some point this happens at a time when conception is likely to occur. Concealed ovulation may be one of the important drivers for long-term relationships, some researchers have argued.

The second issue is how to identify women who are likely to be ovulating regularly, who are fertile and evolutionarily fit. Women after all do not display their fertility directly. However, there are some clues that a woman is more likely to be fertile. The first is physical appearance – 'beauty'. Fertility is associated with youth, and with good female health, and these features underlie our perception of what beauty is: facial features associated with oestrogen production like full lips, high cheekbones, and thin jaws; clarity of skin; length and quality of hair; a low waist-to-hip ratio; firm breasts.[22]

Youth is a strong card in the mating game. While men on average marry women three years younger the first time round, with any successive marriages this age gap widens – up to eight years by the third marriage. In 1800s' Sweden, church records show that, on average, divorced men married women more than a decade younger the second time round.

The most stark modern-age illustration of this tendency for men to favour younger women as partners comes from South Korea.[23] In the early 1990s, with normalisation of diplomatic relations between China and South Korea, a growing number of Korean men, usually elderly bachelors from rural areas, began to marry foreign women. Initially these were Chinese women of Korean ancestry, but with time these men diversified their interest into other developing countries in South, South East and Central Asia. Essentially these brides were bought. The man would pay a matchmaker an initial fee and would be presented with a choice of potential brides. Basic information would be exchanged, regarding age, health and socioeconomic status, although often the matchmakers would be grossly dishonest in the information supplied to the potential bride. Selection would be accompanied by a further fee paid to the matchmaker and to the future bride's family, before a single trip out to visit the prospective bride would seal the deal. These purchased brides would often come to South Korea to find that their new husband was significantly poorer than they had thought, and would frequently be victims of chronic domestic violence.

This trade in brides was so widespread and problematic that the South Korean government legislated to stem the flow in 2014, but in the preceding four years, some 45,000 bought brides had been imported into the country. Comparing over a million Korean–Korean marriages against those marriages presumed to be of purchased brides shows something staggering. In those marriages between South Korean nationals, the bride was fairly consistently a few years younger than the groom. But for these purchased brides, the bride's age did not correspond at all with the groom's age, never rising above twenty-five years old. When women have no limiting role on long-term relationships, essentially stripped of power in the selection of their partner, and men's choice of bride is based upon restricted information, men will prioritise youth above all else, since youth is a proxy for fertility, and those physical attributes associated with it: 'beauty'.

How to avoid raising another man's child

When it comes to long-term relationships, in an evolutionary context, men want a genetically healthy and fertile partner who can bear lots of healthy children. Perhaps equally important as the production of one's own offspring, though, is the avoidance of expending energy and resources unknowingly raising the offspring of someone else. For women, this is not an issue, but as we have seen, non-paternity can affect as many as 4 per cent of all children. It would make sense, therefore, for men to evolve certain patterns of behaviour or preference to minimise the chances of unknowingly raising another man's child. The first adaptation might be a mate preference for virginity, as a predictor of sexual fidelity at the time of commitment and afterwards. However, the evidence of this is poor as an intrinsic human trait, although there are clearly cultural drivers of virginity being a highly valued commodity.[24]

More convincing is the consistent finding that men rate sexual fidelity as one of the top desired qualities in a long-term partner. Since past behaviour is often a predictor of future behaviour, and having a large number of sex partners prior to marriage is a predictor of infidelity in marriage, men tend to evaluate past sexual activity in a prospective partner prior to commitment.[25] Finally, the evolution of male sexual jealousy makes sense from a biological standpoint, guarding against the possibility that a long-term partner may have been impregnated by someone else, as we will see in a later chapter.

Sexual strategies theory argues that commonalities and differences between the sexes when it comes to sexual behaviour have been moulded by evolutionary pressures, all driven to maximise the chances of successfully passing on our genes. An alternative explanation, however, is that some of these differences in mating behaviours stem from physical disparities between men and women, that our physical make-up has led to the development of social structures and gender-based socialisation that define these contrasting mating behaviours. Rather than evolution driving how we choose to mate, it is our societies and social norms, derived from our physical differences as men and women.

In the modern world, in the era of contraception, sex and procreation have become dissociated, and it is not unexpected that societal norms and sexual behaviour have shifted. For the vast majority of human history and evolution, however, this disconnect has not been the case. If sexual strategies theory is indeed correct, there are clear evolutionary advantages, both as a man and a woman, to not sowing your seeds with absolute abandon. Humans have developed the psychological mechanisms, driven by evolutionary pressures, to go down a different route from the majority of animals; to usually prioritise long-term mating over short-term sexual behaviours. Undoubtedly, in part this is influenced by societal norms, the strictures of what is socially acceptable and what is not. Of course, religion influences these views of what is an appropriate way to behave, but frames this in a moral and theological context, reinforcing the rules that we are expected to adhere to. It is nonetheless obvious that our views, and indeed actions, related to sex are mirrored across a range of societies and religions – from the

Hadza in the northern Tanzanian savannah to the multicultural and multireligious streets of London. It is likely that these religious and cultural strictures have their roots in our evolutionary imperatives anyway.

If our psychology is defined by our evolution, then it makes sense that our brains would develop to support those psychological processes. In other words, our neurobiology encourages the inhibition of our basic instincts, to suppress the drive to mate with anyone or anything. Areas of the brain responsible for rationalisation, for quashing our primitive drivers of behaviour, rein back our 'lust'. And when those neurological or psychological systems go awry, lust runs riot.

The brain's basis of our 'sexual brake'

If we suppose that a failure of the evolutionary imperative to suppress or regulate our intrinsic desires emanates from the brain, as in the invalided soldiers we met earlier in this chapter, then how does this happen?

From the earliest days of clinical neurology, from cases like Phineas Gage described in the first chapter, we have known that one of the roles of the frontal lobes – the areas of the brain that sit behind the forehead and above the eye sockets – is the inhibition of socially unacceptable behaviours. Their damage or destruction may result in inability to mask what should otherwise go unsaid or undone. When discussing the three war veterans, Jarvie speculated that their backgrounds and pre-morbid states were either excessively repressed or that, in the second case at least, excessive inhibition of an underlying tendency was required to maintain a veneer of social respectability. The neurobiological restraints of these sorts of behaviours, the endogenous checks on impropriety, had been disrupted by massive brain injury, the normally suppressed 'sexual animal' within unleashed. And the common site of those injuries in Jarvie's patients? The frontal lobes – and in particular the ventromedial pre-frontal cortex (vmPFC) (Figure 3). As with its role in inhibiting anger and aggression, this region of the brain acts as a 'sexual brake' as well. Individuals with damage to the pre-frontal cortex often exhibit a wide array of abnormal behaviours, often with both hypersexuality and aggression in tandem.

This brain region, and its dampening influence on our basic behaviours, is the neurobiological expression of those evolutionary pressures to control our 'lust', to limit who we procreate with.

Jarvie's cases are not the only source to support this view, that the frontal lobes of the brain act as the reins for our sexual desire. A more recent paper also describes a number of patients in whom other types of brain damage were the explanation for their florid hypersexual states.[26] These patients, mostly from Los Angeles, exhibited public masturbation, and attempts to have intercourse with partners and strangers alike, after a variety of brain conditions affected them. In one case, the discovery of a benign tumour affecting the frontal lobes, and subsequent surgery, resulted in a transformation. Initially after the surgical procedure, the fifty-nine-year-old man began to demand sex three or four times a day, with intercourse frequently lasting over an hour due to difficulties achieving orgasm. After nine months or so, his exhausted partner of several years left him. Over the course of the subsequent two years, he became increasingly obsessed with sex. He began to proposition both female and male patients (he had previously had sexual contact with men), and would masturbate on the wards. A scan revealed that he had had strokes within the frontal lobes, a consequence of his difficult surgery.

The neurological origins of lust

While it seems that the frontal lobes, in particular the pre-frontal cortex, restrict the expression of these hard-wired sexual impulses, these regions of the brain are not necessarily the engines of sexual desire – for Simon, for the soldiers, for us all.

In that same Los Angeles cohort, another patient, a thirty-one-year-old woman, had been admitted to hospital with a severe headache. A scan had shown a bleed from an aneurysm, an abnormal outpouching of one of the arteries in her skull. The aneurysm had been surgically repaired, but unfortunately, five days after the operation, she had had a stroke, leaving her weak and numb down the left side. But it was her behavioural change that was most marked. Described as previously a 'shy person', her doctors reported that she had begun to 'talk incessantly about sexual matters and propositioned her physicians. She developed a sexual preoccupation with her internist [physician] and discussed openly in explicit sexual language with him, and others, her desires for him.' She would approach male staff, patients and visitors, without discrimination. Her doctors reported that: 'On one occasion she requested intercourse with a cachectic [weak and wasted] seventy-year-old patient dying of cancer.' Throughout her hospital stay, 'she described intense sexual excitement, and in retrospect she remembers feeling "warm" and "aroused" almost the entire month. In addition to the sexual excitement, she had increased appetite, disturbed sleep, and a verbal output that approached flight of ideas.'[27]

This time, however, in contrast to the patients described previously, the damage to the brain was not in the frontal lobes or anywhere near them. A scan showed a stroke in regions called the thalamus and hypothalamus, deep in the centre of the brain (Figure 1). The hypothalamus will be familiar, since it has already been implicated in Chapter 2 as the core driver of our appetite. Fortunately, within a month, the woman made a gradual recovery and this heightened sexual sensation she described disappeared.

This unfortunate young woman's case demonstrates that areas of the brain other than the frontal lobes might also be involved in influencing sexual behaviour. While the frontal lobes inhibit our behaviours, they are not the neurobiological origin of our sexual urges.

Indeed, another case from the same paper showed that changes in areas of the brain very distant from the frontal lobes can also result in abnormalities in sexual behaviour. One thirty-year-old man had developed a viral infection of his brain, causing damage to his temporal lobes on both sides. The temporal lobes sit on the side of the brain, somewhere between the temple and the ear. After recovery, he had largely lost all interest in sex: 'His wife complained that he never approached her sexually, and their frequency of intercourse dropped from two to three times per week to one or two times per year.' And those rare events happened in very particular circumstances. For this man began to have seizures – typically limited to a small area of his brain rather than full-blown convulsions associated with seizure activity in the whole of his brain. He would often hear sweet, unrecognisable music in the prelude to these seizures, consistent with the origin being in areas specialised for hearing, i.e. the temporal lobe. He would go on to have little recollection for a while after these events. But in the aftermath of these seizures, he would become highly sexually aroused for anything from ten minutes up to twelve hours. According to his doctors, 'his wife became aware of his post-ictal [following the seizure] change and sought sexual relations with him during these periods'. He rarely remembered these occasions afterwards though.

* * *

These individuals illustrate the role of other brain regions, beyond the frontal lobes, in the regulation of our sexual urges. Indeed, damage to areas outside the frontal lobes not only results in heightened sexual desire (hypersexuality); it may also fundamentally change sexual preferences.

In the Los Angeles cohort, the authors described several other individuals with brain conditions that caused a dramatic shift in sexual preference, not just sexual drive. One man, with a healthy sex life, happily married for thirty years, began propositioning his seven-year-old daughter and her friends, and started making sexual advances in public to young children. A scan found what looked like a tumour in the region of the hypothalamus, confirmed on autopsy after his death. Two other individuals, both with infection of the brain by herpes simplex, the virus that causes cold sores which has a predilection for damaging the temporal lobes, also showed huge changes in their sexual preferences. One, a seventy-five-year-old man, began to ask his wife to have sex with other men while he watched. Another, a thirty-one-year-old woman, began to ignore her husband, making 'oral and manual advances' to female members of staff on the hospital wards, with no ensuing recovery.

In contrast to those cases simply exhibiting an increase in sexual activity, where the lack of inhibition in sexual matters was often accompanied by other changes in behaviour, such as excessive eating, this latter group of patients developed an obvious change from a previously stable pattern of sexual behaviour. A heterosexual woman suddenly becoming disinterested in her husband and sexually attracted to women. A man not previously sexually attracted to children, transformed into a paedophile. No single area of the brain was identified as causative of these shifts in sexual preference, but the authors concluded that in all these cases abnormalities were determined in regions in or near a specific system within the brain – the limbic system.

While this network of brain areas is generally considered as being the seat of our emotions, this system also appears crucial to various aspects of our sexuality, such as arousal and choice of sexual partner. The tiny hypothalamus, which constitutes a key component of the limbic system, is not only implicated in our appetite for food, but also in our appetite for sex. As we have seen, this minuscule ancient area of the brain is vital for many of our primitive instincts: the drive to drink, sleep and seek pleasure, not only from food and sex. The hypothalamus can be considered the source of our most basic impulses.

These cases are not the sole evidence for the role of these brain regions in our sexual behaviour. In rats, electrical stimulation of a region of the hypothalamus called the pre-optic nucleus leads them to exhibit copulation, and destruction of this area leads to complete and permanent loss of sexual drive. In humans, electrical stimulation of regions of this circuitry, either by epileptic seizures or direct stimulation during neurosurgery, in particular the areas of the temporal lobe that represent constituents of this limbic system, can result in sexual arousal and orgasm.[28]

The relevance of all of this to Simon's case may not yet be immediately clear, since he has not had a stroke, a bleed or brain injury. Despite his Parkinson's disease, his brain scan is normal. These patients do nevertheless firmly show that changes within the brain can result in rather marked alterations in sexuality, in terms of both intensity and nature.

These clinical cases represent the most extreme examples, where devastating illnesses have left an indelible mark on brain structure and function. Beyond these specific clinical cases, however, to what extent does brain disease intersect with these kinds of 'abnormal' behaviours? For people arrested for exhi-bitionism, a high proportion – up to 35 per cent – are found to have disorders of the brain thought to at least contribute.[29]

For those arrested for paedophilic crimes, about 14 per cent have clear evidence of neurological damage or dysfunction.[30]

But these cases are still at the margins, where glaring abnormalities in brain structure or function give rise to alterations in sexual behaviour of a highly destructive or criminal nature. What significance does that have for the rest of us, those with apparently normal brains, unaffected by damage or disease?

Sex, gender and the undamaged brain

Even for those of us without shrapnel, stroke or other manner of disruption to our brains, it is nevertheless clear that the structure and functions of our brains play a fundamental role in defining our sexual characteristics. What defines our sexual identity and orientation is to some extent predetermined primarily by our genetics – the randomness of inheriting an X chromosome or a Y chromosome from the sperm that fertilised the egg that resulted in us – but not exclusively so.

The chemical milieu in which we grow as foetuses is largely influenced by our genes. Defined by the chromosomes we are dealt with at conception, our testes and ovaries develop in the sixth week of pregnancy. At this stage, however, once our internal sexual organs have developed under genetic control, the production of testosterone by the testes is essential for the development of male external (as opposed to internal) genitalia; the absence of testosterone results in development of a womb and other female genitalia.[31] Being a woman is almost the default state of humanity, only deviated from in the presence of a single chemical. Men are defined by the presence of high levels of that hormone at these crucial stages of our development.

Once sex organs have developed, it is the brain's turn. The balance between male and female sex hormones orchestrates the development of brain circuitry that will go on to influence our behaviours, our preferences, our identities, only for those circuits to be activated by the subsequent surge in hormone levels at puberty.

Our chromosomes, whether we acquire an X and a Y or two Xs, are not the only denominator of the hormonal soup we inhabit as we develop. Other factors can also play a part in determining our gender identity. We have known of this since the late 1950s, when unborn genetically female guinea pigs were exposed to testosterone while their mothers were pregnant. Under the influence of this hormone in utero, these female guinea pigs developed both male genitalia and male mating behaviours, therefore determining both physical and behavioural aspects of gender and sexuality.[32]

In humans, rare gene mutations or chromosomal abnormalities may lead to discordance between genetic sex – i.e. whether an individual carries an X and Y chromosome or two Xs – and physical sex. These genetic abnormalities often affect the production or detection of sex hormones, and thus intrauterine exposure to these chemicals. Even in the absence of these rare genetic mutations, occasionally extrinsic factors may influence our exposure to these hormones. Phenobarbital and phenytoin, old antiepileptic drugs, change the metabolism of sex hormones, and taken in pregnancy, appear to increase the likelihood of 'transsexuality' in the child.*[33] These changes in the developing brain's exposure to sex hormones appear to influence brain structure.

Despite historically some researchers arguing that we are a blank slate at birth, forced by society to take up a male or female gender by external factors, evidence points to gender identity

---

* In this paper, the authors reported higher rates of sex reassignment surgery, but also gender dysphoria, associated with exposure to these drugs in foetal development.

being largely defined before birth. The 'John/Joan/John' case is a good example. Healthy identical twin boys, Bruce and Brian Reimer, were born in Winnipeg, Canada, on 22 August 1965. In the first few months of life though, the boys were noted by their parents to have difficulty urinating. Circumcision was recommended and the infants were referred to the local hospital. Bruce went into the operating theatre first. His surgery was catastrophic, however: the doctors used electrocauterisation (using an electric current to apply heat to cut tissue) and the equipment malfunctioned, burning Bruce's penis beyond surgical repair. His brother Brian's circumcision was cancelled.

The Reimer parents, left traumatised and in limbo for several months, were watching television when they happened to see one of the foremost sexologists of the time, Dr John Money, feature in a debate on sex change operations.[34] In early 1967, they took Bruce to Johns Hopkins Hospital in Baltimore to see Money, who was a strong proponent of the theory of 'gender neutrality'. Money believed that gender identity was largely a result of social learning in early childhood rather than 'predestined' at birth, and perhaps saw a golden opportunity to prove this theory: identical twins, one to be brought up male, the other female. Upon his recommendation, Bruce was castrated and, in infancy, had a rudimentary vulva surgically constructed. He was brought up as a girl named Brenda, given psychological therapy and flooded with oestrogens in puberty. Brenda was not told she had been born male.

According to Money, 'John/Joan' (the pseudonyms used for Bruce/Brenda in his medical papers) grew up to be a happy, healthy woman, clearly confirming his own views that our gender identity is not fixed at birth but can be moulded. The reality for this young person, however, was somewhat different, and much sadder. By the age of thirteen, Brenda was experiencing profound depression and suicidal thoughts, and was told the truth of her past the following year. Brenda decided to assume a male identity, acquiring the name David, and subsequently underwent hormonal and surgical treatments to transition back to male. Contrary to Money's account, some reports suggest that even while living as a girl, Brenda was ostracised and bullied by her peers, and never identified as female. 'She was very rebellious,' her mother later said. 'She was very masculine, and I could not persuade her to do anything feminine. Brenda had almost no friends growing up. Everybody ridiculed her, calling her cavewoman. She was a very lonely girl.'[35]

David later married at the age of twenty-five, and became step-father to three children,[36] but experienced severe mental health problems. By his mid-thirties, he was jobless and separated from his wife, and his twin brother had died of a drug overdose. In 2004, aged thirty-eight, David took his own life.

It is difficult and foolish to extrapolate too much from a single case. There are, however, differences between male and female brains at a very young age. Even by the age of two years old, there is a difference in brain weight between males and females.* On a microscopic level, there are areas that show subtle changes too. The structure that shows most differences is a tiny cluster of neurones deep within the centre of the brain – uncatchily called 'the sexually dimorphic nucleus of the pre-optic area' or SDN (Figure 4). It is not coincidental that the SDN sits within the hypothalamus, that region of the limbic system implicated in some cases of people changing their sexual preference through its damage, as in those patients from the Los Angeles cohort. In men, this region is 2.5 times bigger, with 2.2 times as many cells.[37]

---

* It is important to note that this difference in brain weight does not in any way correlate to difference in intelligence.

Gender identity is complex and is undoubtedly not solely defined by brain structure. However, these genetic and hormonal factors may ultimately influence brain development, and indeed some limited preliminary studies have shown that the tiny nub of brain tissue deep in the hypothalamus – the SDN – is more like the female size in trans-females (male-to-female), and more like the male size in the one trans male (female-to-male) studied.

These types of research are difficult, as they rely on the deaths and brain donations of the individuals concerned. Critics point out that these are very small samples, based on just a tiny number of human brains, and that some of these individuals had been on long-term hormone replacement therapy, which might fuel brain changes. Furthermore, they question how this small nucleus of cells might actually influence our conscious awareness of what gender we feel, independent of our genetic or physical sex. They propose an alternative hypothesis: that these nuclei influence 'male- or female-like' behaviours, and individuals exhibiting these behaviours are viewed as being more male or more female; that individuals become conscious of how they are perceived, and begin to perceive themselves in the same vein. This would imply that gender identity, while being biologically influenced, is not innate.[38]

It is important to stress that these possible structural brain differences underlying transsexuality,* and the view that societal or psychological factors are playing a role, are not mutually exclusive. This is a true ideological battlefield, and research in this area is politically sensitive. However, as with pretty much every area of human behaviour, function, anatomy and disease, there is likely to be a spectrum, with varying contributors – genetic, hormonal, neurological and psychological – for each individual when it comes to gender identity or dysphoria.

---

* The term transsexuality is used here in a medical context, and is quite broad, referring to gender dysphoria and those undergoing gender reassignment surgery and/or hormonal therapy.

Sexual preferences and the brain

When it comes to 'lust' of course, sexual orientation or preference is sometimes entirely separate from our gender identity. What then is the evidence for our sexual orientation, not just our gender identity, being influenced or even defined by our brains?

Despite many past and present attempts to get people to change their orientation for societal or religious reasons – castration, hormonal therapy, psychoanalysis, inducing vomiting while looking at homoerotic pictures, brain surgery, electroconvulsive therapy – it is surprisingly difficult, if not frankly impossible. This is probably one of the strongest arguments against the sometimes expressed view that homosexuality is a 'lifestyle choice', or in some way facilitated or caused by our social environment. It is argued that our sexual orientation is essentially fixed by the time we reach adulthood. Exceptions include damage to the brain in regions like the hypothalamus or other aspects of the limbic system, as in some of the cases we've looked at. Even in other species, such as rats or ferrets, experimental destruction of the pre-optic area can induce change in sexual orientation.

So, if our sexual preference is defined by our brains, is there evidence of brain differences according to sexual orientation? When researchers have compared the brains of homosexual and heterosexual men, they have indeed found differences in various regions of the hypothalamus or in tracts that communicate between the temporal lobes of either side of the brain. These differences are not just seen when looking at slices of the brain down a microscope, in people that have died. Modern imaging techniques permit us to look at the functioning brain, in living people. Several studies have demonstrated differences in how the hypothalamus functions, either in response to certain drugs,[39] to pheromones[40] or to erotic videos.[41] Further research based upon a technique called volumetric analysis – essentially using MRI to provide an objective three-dimensional assessment of brain volume – has also found some differences in symmetry between the two sides of the brain. Heterosexual women and homosexual men appear to have more symmetrical brains, while homosexual women and heterosexual men will have larger right hemispheres than left.*42

Parkinson's disease, hypersexuality and dopamine

Evidence from both animals and humans therefore implicates the structure and function of our brains, moulded by our genetics and exposure to chemicals in the womb, in how we live our sex lives, our sexual behaviour and our orientation. These regions of the brain are not just associated with these behaviours. They directly influence them, as shown by experiments or cases where these areas have been damaged or disturbed in some way.

In real life, however, this reductionist perspective, of ascribing our sexual behaviour to changes detectable on brain scans, is rarely so straightforward. Simon's case illustrates this starkly.

Unlike the historic cases detailed before, Simon has no major structural damage to his brain: his scans are normal. There is no stroke, no bullet or shrapnel that has penetrated his skull. He does of course have a degenerative condition of the brain in the form of Parkinson's disease, but his sexual behaviour began to change several years before he was given a diagnosis.

A change in sexual preference or sexual behaviour may not always be due to alterations in brain structure, however. It may

---

* In the same studies, there were differences in connectivity between other regions of the limbic system, including connections to the amygdala and the anterior cingulate.

also be due to the unchanged brain structure functioning differently. A clue comes from the Los Angeles cohort of patients described earlier and may be of particular relevance to Simon. Among the individuals with strokes, bleeds, tumours and viral infections of the brain, one patient in particular stands out: a seventy-one-year-old man, diagnosed with Parkinson's some ten years previously. He had recently been started on treatment, which helped his tremor, his paucity of movement and stiffness. But on the drug, the man's behaviour changed completely:

> He became preoccupied with intrusive sexual fantasies and ruminations concerning sexual topics. He began to insert objects into his penis and on one occasion a pencil had to be surgically removed. His medication was reduced and the aberrant sexual behaviour abated. He had no history of psychiatric disturbances or previous sexual dysfunction or paraphilic behaviour. He had been married for over thirty years and his wife also denied that he had ever shown an interest in atypical sexual practices.[43]

The drug that he had been commenced on, that had induced his dramatic alteration of sexual behaviour, termed a paraphilia, is called levodopa. This amino acid, once it crosses into the central nervous system, is converted by an enzyme into a familiar chemical: the neurotransmitter dopamine. In Parkinson's disease, the major hallmark is the loss of dopamine-producing neurones in an area of the mid-brain called the substantia nigra ('black substance' in Latin – these dopamine-producing neurones look darker than the surrounding areas to the naked eye), part of a network of nuclei called the basal ganglia. Levodopa aims to normalise dopamine levels in this region, thus restoring a degree of motor function. The drug, first used in humans in the early 1960s, still remains the mainstay of treatment for this neurological disorder.

As early as 1969, however, reports began to emerge of hyper-sexuality precipitated by the use of levodopa in previously sexually 'normal' individuals with Parkinson's disease. That initial paper reported on the author's experience of treating eighty Parkinson's patients with high doses of oral levodopa, and demonstrated huge improvements in their motor function. But it also detailed problems with side effects. Among these eighty patients, a number exhibited marked personality changes:

> A behaviour pattern with frontal lobe overtones was evident in some patients, especially in the sexual sphere. A clear-cut, visually evident increase in libido occurred in at least four male patients but, unfortunately, erections were not sustained and copulation was terminated with premature ejaculation. It is difficult to evaluate the presence or absence of this effect in our female patients, but we believe that it is present in them as well.[44]

Since that first report, it has become more widely reported that hypersexuality is associated with the use of levodopa, and more modern drugs too – dopamine receptor agonists, that stimulate the dopamine receptors in the brain. Three per cent of patients exhibit hypersexuality on these medications,[45] typically manifesting as increased demands for sex from their partners, an excessive interest in pornography or compulsive masturbation. Nowadays, I routinely ask all my patients living with Parkinson's disease if they have noted anything like this, which can make for some uncomfortable conversations. At the time, they often do not realise that anything is wrong, that their behaviour is unusual. It is only when the drugs are stopped, and the hypersexual behaviour lessens, that the implications of their actions dawn on them.

In recent years, we have learnt that these medicines not only trigger hypersexuality, but can also dramatically alter sexual preferences, as in the man described above who began to insert objects into his penis. In fact, the range of paraphilias (defined as intense urges or behaviours involving 'non-normative' sexual interests) caused by these drugs is absolutely mind-boggling,[46] and includes exhibitionism, frotteurism, paedophilia, sexual masochism, transvestitism, voyeurism, telephone scatology, zoophilia and klismaphilia (arousal from enemas).

Occasionally, with these drugs, multiple types of paraphilia have been seen in a single patient. One individual was found to have developed exhibitionism soon after starting levodopa, and was charged with indecent exposure, before then starting to cross-dress, make obscene phone calls and follow women down the street.[47] Another patient, taking higher doses than prescribed of a similar drug (one that binds to dopamine receptors), developed hypersexuality and was discovered by one of his sons attempting to have intercourse with the family dog.[48] Eight of the thirty-one patients with paraphilias described in one review faced criminal consequences as a result of their actions.[49]

While most patients in the medical literature are males, it can affect females as well. And in all cases, reducing dopamine stimulation through lowering the doses of these drugs or providing another drug to block dopamine – essentially counteracting the pharmacological effects of levodopa and similar agents – caused an improvement or resolution of these paraphilias.

So how can a simple amino acid, a precursor of dopamine, cause such a dramatic and awful transformation in someone's character, leading to distress, harm of self and others, even imprisonment?

Besides dopamine's role in the regulation of movement, we have already heard of its role in the neurobiology of pleasure and reward when it comes to gluttony. In fact, these sorts of drugs have also been linked to compulsive eating. But dopamine's role is not just limited to the reward of food. This pathway of pleasure – those dopamine-producing neurones that constitute the mesolimbic pathway that projects to the 'reward centre' of the nucleus accumbens – is the mediator of pleasure and reward when it comes to sex too (Figure 5).

From the perspective of hypersexuality and paraphilias therefore, dopamine, and the pathways it mediates, seem fundamental to what makes us feel good, what gives us pleasure, what drives us, and what we seek out. It might thus be expected that messing with our dopamine levels might alter our response to positive stimuli such as sex, and make us seek it out more, sometimes to extremes. And it may also drive us in novelty-seeking behaviours, finding new experiences that bring reward or pleasure, perhaps associated with excessive risk-taking in the search for such pleasures.

Precisely how this 'non-normative' sexual behaviour arises remains uncertain. One potential explanation is that intermittent exposure to high doses of dopamine – by taking, for instance, large doses of levodopa three or four times a day – in the nucleus accumbens results in this nub of brain tissue becoming much more sensitive to the effects of dopamine, a 'hypersensitisation' to these chemical effects. This increased sensitivity is what drives the hypersexuality and paraphilias. It amplifies the craving for pleasure in heightened and novel forms.

Dopamine and sex addiction

In light of all these links between Parkinson's disease, dopamine and hypersexuality, perhaps this is the explanation for Simon's actions. Except for one glaring issue. He has only very recently been started on treatment for his Parkinson's, and his change in sexual behaviour started several years before his diagnosis. We therefore need to look for an alternative explanation.

We are not quite done with dopamine though. Because this little chemical may be of relevance to the abnormal sexual behaviour of people without brain damage or the effects of external chemicals like medications. While we are accustomed to headlines splashed across the tabloids featuring celebrities like Michael Douglas, Charlie Sheen and Tiger Woods, who have announced themselves as sex addicts, this is not just a disorder of the famous; approximately 6 per cent of the adult population have features of sex addiction[50] (one study has estimated that 15 per cent of Christian clergy met criteria for cybersex addiction[51]). The nature of sexual behaviour ranges from problematic use of pornography to hypersexuality, and in extreme cases, to paraphilias as well.

Not all psychiatrists or psychologists believe in the concept of sex addiction as a disorder. Indeed, it is not listed as one in the latest edition of the *Diagnostic and Statistical Manual of Mental Disorders (DSM)*, published by the American Psychiatric Association and viewed as one of the diagnostic bibles in the world of psychiatry.[52] Some clinicians argue that sex is a normal behaviour, associated with good health and well-being, and to consider this an addiction is incorrect. Rather, they propose that it is a condition of impulse control, a failure of suppression of basic urges.[53] The argument is not whether compulsive sexual behaviour exists, but whether it represents addiction in the same way as substance addiction, or simply represents difficulty in the regulation and inhibition of normal impulses. But there are also arguments about how to diagnose it, and how not to medicalise people.

Regardless of the classification of this disorder, there are clear overlaps with other types of addiction. There is the loss of control, the continuation of hypersexual behaviour despite significant harms to oneself or others, and an obsession or preoccupation. It is not the sex itself that is the problem (be that real sex, fantasy or pornography). It is the consequences of it, and that people wish to be free of it.

The majority, about 80 per cent, of sex addicts are male, and these men are much more likely to indulge in paying for sex, pornography and paraphilias. Some researchers suggest that women are more likely to exhibit aspects of 'love' addiction rather than purely sex.[54] Many sex addicts also have substance abuse issues, or find that drug use may precipitate a relapse of their compulsive sexual behaviour. Whether this reflects a brain more prone to addiction, or that many drugs increase desire and may precipitate sex addiction, is uncertain.[55]

Like drug addicts and those with abnormal appetites for food, there are parallels. Some sex addicts describe a 'wanting' of sex rather than a 'liking' of it. Occasionally they report an enjoyment of sex with their regular partner that is simply not present during their extra-marital sexual pursuits.[56]

While the argument over whether sex addiction is a true addiction or an impulse control disorder (or even something else) continues, the same circuitry implicated in those individuals with Parkinson's disease is at play in these people too. That pathway of pleasure – the mesolimbic pathway – underlies wanting but not necessarily liking.

Functional brain studies – where brain activity rather than structure is visualised – provide important clues. In men with problematic pornography use, a wide range of regions involved in the mesolimbic circuits react more vigorously to sexual prompts.[57] Even cues that are predictive of the display of erotic images, not the erotic images themselves, cause the nucleus accumbens to light up in these individuals. In fact, when it comes to actual response to the erotic images themselves, there are no differences between those men with problematic pornography use and those without, nor do they rate the liking of those images any differently. This implies that the problem of sex addiction is primarily caused by an increased motivation to seek out these sexual stimuli, rather than due to the enjoyment that such experiences bring. It also suggests that the circuitry that underlies this wanting has been augmented in some way, sensitised by repeated stimulation.

In people with Parkinson's disease, it is thought that it is external dopamine itself, given as a precursor drug, that intermittently boosts levels beyond what would be expected in normal physiological circumstances. But in these sex addicts, what is causing this sensitisation? What is the external substance that is causing stimulation of this circuitry?

One proposed explanation is that our brains did not evolve to exist in today's hypersexualised world, surrounded by easy accessibility of erotic or pornographic images, of readily available sex at the tap of a touchpad or at the end of a phone app. That this modern sexualised world represents an environment that we are not designed for, that predisposes to excessive stimulation of this circuitry, and puts those vulnerable, due to their genetics, brain chemistry or psychology, at risk of these destructive behaviours. And truly destructive they can be, for individuals and those around them, with a host of physical, psychological and legal consequences – from the abuse of drugs to enhance or facilitate sexual performance, contracting sexually transmitted diseases, cosmetic procedures to enhance appearance, to shame, guilt, unplanned pregnancies, marital breakdown, arrest or imprisonment for sexual assault, or job loss due to acting out in the workplace or focus on things other than work.[58]

If indeed this is correct, that our sexual environment is the cause of this accentuation of brain circuits that promote the search for more, then perhaps it is a potent argument for a regulation of societal norms, whether by religion or law. If exposure to sexual imagery may drive or at least contribute to some of these harmful behaviours, then limiting or restricting our environment may ultimately benefit the common good. And as we have already heard, our environment, particularly in childhood, may fundamentally alter the function and structure of our brains.

So, is it simply the case that Simon is a sex addict, that his troubles have nothing to do with his Parkinson's disease? There are certainly some aspects of his story that would support this view – his preoccupations, his anticipation of sex rather than the enjoyment of it, for example. Maybe the preceding use of pornography was the source of the hyperstimulation of his nucleus accumbens. However, equally, there are some features in what he says that point in another direction entirely.

In the aftermath of his confession to his partner, not only did he lose her and his daughter, but also his friends. When his partner informed them of what had happened, his entire friendship group cut him off. 'They just stopped talking to me, because they thought what I did was so unforgivable.' He remains in vague touch with one of his friends, but apart from this, and shared custody of his daughter, he is isolated and lonely, his life pulled apart. 'I was living very comfortably with my partner and child, and had a really good friendship group. And now all that is gone.'[59]

Curiously, Simon has not mentioned any of this to his neurologist, who looks after his Parkinson's disease, due to embarrassment and shame. Nor has he discussed this with a psychologist. I am intrigued as to why he is telling me all this now. I get the impression that he is in the midst of profound introspection. Simon is desperately keen to understand himself and his actions, in the context of his deep mortification, a year on from the devastation that this has wreaked upon his and his family's lives. He is worried that unless he recognises and addresses it, this will be an ongoing problem in his life, destined to destroy his future as well as his present.

Simon's case is complicated, made more so in the setting of his Parkinson's disease. Could his neurological disorder be related to his sexual behaviour? Based on the chronology, the onset of his hypersexuality many years before commencing on levodopa, we can rule out the effects of medication. Rarely, these issues can be seen in patients with Parkinson's disease without medication, but Simon has not experienced any other impulse control problems – no gambling, no addictions, no eating issues, no excessive shopping or spending. In fact, since starting on levodopa, the drug has improved his psychological status, not worsened it: 'It has made me feel a little less sad.' Nor have his sexual behaviours had any impulsive qualities to them. They have been preceded by days, weeks or months of rumination, of obsession, with perhaps the exception of the last event, the sleeping with his friend. The majority do not have the flavour of a spur-of-the-moment impulsive act.

Simon recognises that some of this may be driven by feelings of low self-worth, of concerns that women do not want to have sex with him. Nevertheless, the overall picture implies an obsessive-compulsive component, driven by anxiety. Those recurring thoughts of sex, the increasing distress, the stilling of his ruminations with the act. Impulsive and compulsive acts are quite distinct. While impulsivity is characterised by unplanned actions carried out without regard for the consequences, compulsions are different. These are actions undertaken to avoid or resolve inner feelings of distress or discomfort, but do not give rise to positive feelings, of pleasure or reward.* There are

---

* While there is significant overlap between addiction, impulse control disorders and compulsive behaviour, there are key differences. Addiction describes the phenomenon of repeatedly engaging in detrimental behaviour to seek out what they believe to be pleasurable, such as a drug or alcohol 'high'. Impulsivity is defined as unplanned actions or reactions to stimuli, be they internal or external, carried out without regard for the consequences. In contrast, compulsions are actions, often repetitive in nature, undertaken to avoid or resolve inner feelings

also other features that point in this direction too. His obsessional thinking about sex is heightened at points in his life when
other anxieties are also prominent, particularly his health anxieties, and treatment with an antidepressant has lessened both
his anxiety and his sex-related obsessional thoughts. I suspect
that there are also some longstanding tendencies to obsession-
ality anyway – his fixation with computers at an early age, or
his 'fascination' with the woman he met on holiday, with whom
he subsequently embarked on a long-distance relationship.

It is obvious that his personality does not fully explain what
has transpired, however. He reached his mid-twenties without
any compulsive sexual behaviour, maintaining stable long-term
relationships. He has a great deal of emotional empathy for his
ex-partner. He is not a narcissist or a psychopath, able to simply
do what he wants, to seek pleasure without any consideration
for those around him. Rather, he is riven by guilt and embarrassment, seeking answers to why he has done what he has
done.

It may be that alcohol simply lowers his inhibitions, making it
more likely that he will act on his compulsions; alcohol impairs
frontal lobe function, easing off the brakes on our behaviour. As
we have already seen, the nature of sex addiction and behaviours
of that ilk remains controversial, and arguments persist as to
whether it is a true addiction, an impulse control disorder, or could

---

of discomfort or tension, but do not contribute to positive feelings. Instead,
compulsive behaviour often brings with it distress, rather than the positive feelings
associated with addiction and impulse control disorders. In OCD, these compulsions are aften associated with obsessional intrusive thoughts, and the compulsions
are behaviours that seek to reduce these obsessional thoughts and their associated
unpleasantness. Whereas in anxiety, people may worry about realistic eventualities
but their responses to those worries is excessive, in OCD the worries are often
not connected to reality. For example, someone with anxiety may worry if they
are overpaying for their car, and spend excessive time researching car prices, while
someone with OCD may worry that they may crash their car unless they switch
their lights on and off in a particular sequence.

perhaps represent something more akin to obsessive-compulsive disorder, at the extreme end of anxiety. Certainly, compulsive sexual ruminations are seen in OCD, although typically there is overlap with other compulsive behaviours, and it is unusual to act on these sexual thoughts.

So, is the Parkinson's disease a red herring then, completely unrelated? Maybe, but maybe not. Increasingly, it is recognised that before the onset of frank (obvious) features of Parkinson's disease – the tremor, stiffness and walking difficulties – there is a prodrome of subtle symptoms that manifests some ten to twenty years beforehand, as the degenerative changes within the brain slowly develop. These include sleep disturbance, and loss of sense of smell, but also psychiatric symptoms such as depression and anxiety.[60] These are likely to originate from changes in the brainstem and the limbic system, the network of the brain most involved in emotional regulation. It is quite possible, therefore, that the Parkinson's disease incubating in Simon's brain might have played a role, driving some of the anxiety that has fed his obsessional thoughts of sex, and ultimately resulted in his damaged life.

Understanding the nature of his behaviours will not rectify the situation he finds himself in but may well help prevent Simon's life unravelling again in the future. And importantly, it illustrates the interactions between the 'psychological' and the 'neurological', how the interplay between brain and mind influences why we do what we do. I use these terms in inverted commas for a reason. These distinctions are blurred. The division between the mind and brain is grey, perhaps even non-existent.

For as we have already seen, and will continue to do so, our emotions, our feelings, our thoughts, our behaviours, originate in the structure and the function of the brain. As the individuals above show, a change in how our brains work can transform our behaviours. Sometimes, that change is obvious, evident on a scan or on a post-mortem microscope slide. Sometimes, changes are less tangible, more elusive. If a change in brain function can result in a change in our 'lustfulness', however, then the origin of 'lust' is obviously our brains.

Our drive for sex is a biological imperative. Its outward manifestations are simply a result of the machinations of those areas of our brains that drive it and hold it back. It is when it is amplified or unrestricted, through brain injury, medications or our environment, that it ruins lives. But without 'lust', we are destined to no life at all.

4.

Envy

Envy could be seen, eating vipers' meat that fed her venom . . . Her sight is skewed, her teeth are livid with decay, her breast is green with bile, and her tongue suffused with venom. She only smiles at the sight of suffering. She never sleeps, excited by watchful cares. She finds men's successes disagreeable, and pines away at the sight. She gnaws and being gnawed is also her own punishment.

Ovid, *Metamorphoses*

'Sarah was thinking I was having affairs with friends. Friends I have known for thirty-odd years. Sarah thought I was cheating on her. No one was above suspicion – family, friends, everybody. You couldn't rationalise or reason. You could not talk her round at all.' Colin sits next to his wife on the sofa, while she shifts restlessly as he speaks. 'It did get violent on occasion. Sarah venting her frustration, because initially I didn't want to leave. So, I would try to stay to look after [her], but it would escalate and then become violent. I would be aggravating Sarah just by my presence. Because I had been unfaithful, Sarah was ordering me out, because she just couldn't tolerate me in the same house. So, the last time, I went to stay with my daughter, and the twins came too.'

'I feel terribly guilty about all this,' Sarah chips in.

Both now in their late forties, they have been together for twenty-six years, with an adult daughter and eighteen-year-old twin boys. In all their decades together, Sarah had never behaved like this. '[Normally] I wouldn't hurt a fly,' she says. 'That's true,' Colin confirms.

Her memories of these episodes are initially moth-eaten, the holes gradually filled in: 'It came back to me in flashes afterwards. Flashes of remembrance.' On realising what she had done, how she had behaved, she would break down and cry, overcome with guilt. 'I apologised to my friends, family and children. I have apologised a million times.' But at the time, she says she was '100 per cent convinced' that Colin was being unfaithful. Her unshakeable conviction that her husband has been adulterous, even when presented with facts that definitively disprove this, meets the objective clinical threshold of a delusion.

Her delusions of infidelity are elaborate, and any evidence to the contrary she believes to be fabricated. These 'affairs', Colin explains, 'had been going on for decades, essentially. It was all subterfuge, and even people in Sarah's family were involved. They were all helping me. For example, she said I had gone on holiday with one of my female friends, that we had been having this torrid affair. Obviously, we could pinpoint where we had been at that time, and that it did not happen.' But Sarah simply would not believe it. She would say: 'It was false, just a lie, that evidence. It had been orchestrated by multiple people to look that way.'[1]

Three episodes of intense and destructive jealousy have blighted their marriage in the last couple of years, with Sarah admitted to a mental health unit on two occasions, once voluntarily and once forcibly. Intense delusions surrounding sexual infidelity, unshakeable in the face of clear evidence to the contrary. Despite these admissions, psychiatrists have been unable to find obvious features of any ongoing mental health issues. 'They are continuing to investigate,' says Colin.

* * *

Ovid describes Envy as a poisonous figure full of malice towards humankind. A far cry from more modern interpretations of envy that reflect the range of human experience of this emotion. At its core, envy can be defined loosely as the desire of, or the wish to see someone deprived of, the superior qualities, possessions or achievements someone else has. Psychologists often define the desire of those qualities as 'benign' envy, while the wish to deprive someone of their superior characteristics is 'malicious' envy.

Then there is jealousy, often used interchangeably with envy, but not quite the same thing. From a psychological perspective, jealousy involves the threat of someone taking something or someone away from you – like Sarah's pathological jealousy regarding her husband. However, even in the scientific literature these distinctions are not always clear. Some psychologists argue that jealousy more often has a sexual context, or is more possessive, wanting to deprive someone of their possession and claim it for oneself.[2] The emotion of jealousy usually has a profound negative quality to it; envy less definitively so. However, envy and jealousy often coexist. You may feel threatened by a rival precisely because they have qualities or possessions that you envy.

Envy is a hostile emotion, one that is the keystone of crimes, conflicts and biblical tales. Envy resides in the Ten Commandments, carved into stone tablets, carried down from Mount Sinai by Moses. It is the motivation of the second crime mentioned in the Bible, Cain's murder of Abel: 'The Lord looked with favour on Abel and his offering, but on Cain and his offering he did not look with favour. So, Cain was very angry, and his face was downcast.'

This human trait has been proposed by some theologians as the core emotion that drives most sinful acts.[3] Its potency extends beyond the detriment of others and nurtures a willingness to sacrifice one's own position to deprive those around us. It is the emotion that Nietzsche postulated as the basis for the egalitarian moral framework of Christianity, and which philosopher Ayn Rand claimed was emblematic of much of the latter twentieth century, leading to the 'hatred of the good for being good'.[4] It is an emotion that is painful, blended with feelings of inferiority, resentment and negativity directed at our fellow humans.

Nowadays, when I hear people talk of being envious, it is apparent that this term has taken on a very different meaning. The modern view of envy is often one that is almost entirely devoid of hostility. Benign envy verges on admiration, and touches on aspiration. Some psychologists and philosophers point out that this benign envy, 'may obscure the nature of envy. The absence of hostile feelings in benign envy may render the emotion fundamentally different from envy proper both in terms of the felt experience and in terms of its likely consequences . . . In our view, benign envy is envy sanitised . . . and lacks a core ingredient of the emotion, namely some form of ill will.'[5]

All emotions, envy and jealousy included, are drivers of survival. A response to our environment may be rather simple, such as seeing water and drinking it, but emotions drive a more organised and more intricate pattern of behaviour. They influence a wide array of functions, both neurological and physiological. And some emotions are more complex than others. Fear and anger are more intuitive, more clearly triggered, with outcomes that are more immediately obvious and useful from a survival perspective. In contrast, emotions such as envy and jealousy, and indeed others like guilt, shame or pride, require a degree of introspection, some self-knowledge – a linkage of these emotions to higher areas of cognition and a deep understanding of the social world we inhabit. The evolutionary imperative of these simple emotions is also more immediately apparent too, primal guardians of our existence. Fear, for example, reduces the risk of predation or intra-species conflict; anger drives violence in the defence of possessions such as food, partners or territory.

So why is it that we experience these unpleasant emotions: envy and jealousy? Why does someone else's advantage induce these painful feelings? What are the evolutionary benefits, those survival advantages that these emotions bring? The most obvious explanation is that advantages held by others have profound consequences for oneself. Since these advantages confer benefits, in terms of mating, food, success in broader terms, and we live in a competitive world, one person's advantage is another's handicap.

The two faces of envy

Benign envy, the sort without hostility, may have some very positive outcomes. Comparison of oneself to others deemed superior may actually drive feelings of hope – that improvement and achievement are within grasp.[6] It may also drive creativity and motivation. The key here, however, is to envy someone like you, someone who you can emulate, who is within your realm. From an evolutionary perspective, it makes no sense for us to envy Elon Musk's, Jeff Bezos's or Bill Gates's stupendous wealth – their position is so far outside our reach that to envy them serves no useful purpose. In contrast, to envy our neighbour, who earns a little more, or drives a nicer car, has much more evolutionary logic. As the famous neuroscientist V. S. Ramachandran and his colleague write: 'The whole purpose of envy is to motivate you into action either by independently trying harder (envy) or by coveting and stealing what the other has (jealousy) . . . Envy evolved to motivate access to resources that are in demand by others in your group.'[7]

That envy proper, 'malicious' envy, has a hostility associated with it – an unpleasantness – serves as an emotional reinforcer of that motivation, that striving to compete for limited natural resources. Hostility makes one more resolute, more focussed. The alternative would be to be submissive in the face of one's own inferiority, for fear of being harmed by a superior; of reprisal. The hostility associated with envy proper may help one break free from this tendency to submission.

Like other emotions, therefore, envy and jealousy have an evolutionary purpose. They motivate us to better ourselves, to compete for finite resources, to defend what is loved by us from others. These emotions are fundamental to our being, and arguably responsible for the progression of the human race.

Yet it is these very same forces, when out of control, that cause murder and mayhem, conflict and war. I can understand the ancient religious proscriptions, that we should value ourselves by our moral and spiritual virtues rather than external factors. That our self-esteem should not depend upon our wealth, our looks or our possessions. If we find ourselves wanting in our internal sphere, we should try to improve ourselves rather than envying others. But these religious strictures fight against our nature, sometimes to be defeated. And even the religions recognise the complex nature of envy, that it has positive and negative attributes. Jewish texts contain proverbs such as: 'Be envious for my sake! Were it not for envy, the world could not be sustained. No one would plant a vineyard, no one would take a wife, no one would build a house.' Similarly, Islam recognises the difference between *hasad* – malicious envy – and *ghibtah*, benign envy, which leads enviers to work for Allah.[8]

Blurred boundaries between the normal and pathological

Both envy and jealousy can therefore serve us, but may also become pathological. From a clinical perspective, it is when these emotions are so intense that they create distress in everyday life, in relationships – when they result in harmful, sometimes dangerous, conduct – that they come to medical attention. In Sarah's case, her jealousy was so profound that it resulted in violence and led to her being involuntarily admitted to a psychiatric hospital.

In fact, envy is rarely viewed as a medical issue but remains largely in the domain of the philosophical, theological, social and psychological. It is usually in the realm of the normal rather than pathological, although the boundaries between the two are so often indistinct and arbitrary. In the context of envy, this problem of defining pathology is of particular significance when it comes to personality disorders, as we will see.

This issue of what constitutes a state of being that is simply at the extremes of the spectrum of normality, and what tips over into becoming a disease or disorder, riddles the world of medicine, and psychiatry in particular. In the absence of clear tests, on blood or on brain, that hoist a red flag declaring a pattern of behaviour or thinking as being a disease, the diagnosis in most cases rests upon the subjective view of the doctor. It is a matter of interpreting what the patient, or their family, might be telling them.

In many cases, abnormality may be obvious, manifesting as florid hallucinations or delusions – those false beliefs so firmly held despite evidence to the contrary – but frequently it is not. Hence the persistent debates between various schools of psychiatric and psychological thought about how the *Diagnostic and Statistical Manual* (*DSM*) defines and classifies mental health conditions. There are frequent accusations of the medicalisation of normality (often accompanied by the suggestion that the pharmaceutical industry has lent a hand, to fuel prescribing of their wares). These arguments have relevance for many of the Seven Deadly Sins. Even the editor of one of the previous iterations of the *DSM* has heavily criticised the latest fifth edition, *DSM-5*. In an article in *The Huffington Post*, he rallied against the dangers of defining heightened normal behaviours as psychiatric diagnoses, that *DSM-5* risked turning temper tantrums, normal grief or occasional binge-eating into psychiatric disorders. And by labelling these as disorders, it opened the door to people being unnecessarily medicated: 'New diagnoses in psychiatry are more dangerous than new drugs ... *DSM-5* has created a slippery slope ... to make a mental disorder of everything we like to do a lot'.[9]

These arguments among psychiatrists and psychologists regarding what constitutes mental illness are distilled in the concept of personality disorders, and whether these conditions truly represent a mental illness. As we have seen in relation to borderline personality disorder, more broadly, personality disorders are typically defined as an enduring inflexible pattern of thought processes or behaviour that deviates significantly from that expected within someone's culture, leading to distress or impairment in varying aspects of life. These thoughts or actions are not attributable to medications, recreational drugs or another medical or psychiatric disorder, but are innate.

While there are differences between definitions in the various standard classifications such as the *DSM* and the *International*

Classification of Diseases, essentially personality disorders are considered to represent normal personality traits that in that individual are so extreme as to cause harm to themselves or to those around them. By definition, this description recognises that these features of our personalities exist in all of us to a greater or lesser degree, that we all sit somewhere on a spectrum of all of these aspects of human behaviour. The challenge when it comes to these traits, including our propensity to 'sin', is to define the point at which normal becomes abnormal; when a trait becomes a disorder.

As such, personality disorders, like our psychological traits, are at the core of that individual, a function of their longstanding personality, with resulting implications that they are less amenable to treatment and cannot be 'cured'. Many of the terms used by psychiatrists to define or subdivide these personality disorders have entered everyday parlance: histrionic, paranoid, borderline (now more usually termed 'emotionally unstable'), narcissistic, sociopathic, etc. These disorders are incredibly common, with up to 10 per cent of the population meeting diagnostic criteria.[10]

While the origins of personality disorders are poorly understood, they are thought to result from interactions between an underlying neurobiological predisposition – in part genetic – and stresses during development. As we have already seen, childhood trauma, especially abuse or neglect, is a known risk factor.[11] Yet increasingly it is recognised that there are strong biological factors at play. Personality is strongly influenced by hereditary factors, and twin studies comparing identical and non-identical twins suggest that genes have a potent role in the development of personality disorders too.[12] Research in personality disorders also suggests differences in neurotransmitter systems, volumes of certain brain structures important in the regulation of emotions and disruption in networks that are involved in behaviour.[13]

In keeping with the view that these disorders are hard-wired into the brain, and are not as a result of a disease process affecting the mind or the brain, most psychiatrists consider personality disorders not to be mental illnesses.[14] While undoubtedly personality disorders are risk factors for mental illness, and complicate its treatment, they are viewed as distinct, separate entities. You could argue that a personality disorder ultimately reflects a difference in brain function, possibly even structure, in much the same way as any other psychiatric or neurological disorder; that it could be considered in a similar light to other conditions which develop early in childhood and persist throughout life, such as autistic spectrum disorder, for example. And that this is purely a question of semantics: how you define an illness, a disease or disorder. However, it does not change the fact that treating a condition that is enmeshed in the fabric of our brains, that is an intrinsic aspect of who we are, is going to be more challenging than treating an abnormality of the brain resulting from infection, inflammation or chemical imbalance.

Personality and envy

Of particular relevance to envy is one specific personality disorder: narcissistic personality disorder (NPD). At first glance, this seems very counterintuitive.

The hallmarks of this personality trait when extreme, as in NPD, result in a sense of being special: an entitlement, feelings of self-importance to the point of grandiosity, preoccupations of brilliance or success and excessive arrogance. These features are accompanied by a lack of empathy, the tendency to exploit others to achieve their own ends, and attention-seeking behaviour. People with narcissistic traits (rather than the personality disorder) will often be extremely successful, as these characteristics facilitate sociability and getting what you want. But in people with the personality disorder, rather than just having narcissistic traits, their overconfidence and antisocial behaviours – manifesting as looking down on others, impatience when others are perceived to be getting more attention and hostility when criticised – impair their ability to function in the workplace and in their social and family lives.

In the context of heightened self-importance and feelings of superiority, the emotion of envy seems distinctly out of place. Surely you only envy people who have more than you, achieve more than you or are somehow else more successful than you. After all, if you go through life feeling the best, the cleverest, the most beautiful, the most interesting, what do you have to envy? Yet despite this, many people with NPD exhibit very heightened feelings of envy, contributing to the destructive nature of this disorder.

This apparent enigma suggests that narcissists may simultaneously feel terribly superior and painfully inferior to others.[15] There is an explanation for this puzzle, however, since there are different subtypes of NPD. On the one hand, there are those individuals with enormously inflated egos, with marked grandiosity, which presents as an exhibitionistic, domineering and aggressive persona. On the other, there are those individuals who exhibit what is termed narcissistic vulnerability: while vulnerable narcissists share inflated self-importance and entitlement with grandiose narcissists, they also feature hypersensitivity, vulnerability to traumas and shame. They are more likely to be sensitive to rejection, have a distrust of others and increased levels of hostility and anger. They will fear being laughed at – and enjoy laughing at others. Their strong need for admiration and sense of entitlement will drive some of their overt behaviours.

Unsurprisingly, it is these vulnerable narcissists rather than grandiose narcissists who will demonstrate very heightened levels of envy, the latter group protected by their own grandiosity – 'I am the best, I have nothing to be envious of!' Vulnerable narcissists have a particularly toxic combination of a feeling of entitlement and worries about inferiority and shame.[16] And envy also sets the stage for Schadenfreude, an elevated level of pleasure at the misfortune of others, and leads narcissists to sabotage them.

The extremes of jealousy

In stark contrast to envy, jealousy often warrants the attention of the medical profession, as Sarah and her episodes of intense and unexpected jealousy illustrate. In her case, the intensity of this emotion, and its unshakeable nature, meets the definition of morbid or pathological jealousy.

Sarah grew up the youngest of thirteen siblings in the east end of Glasgow. She describes a tough upbringing in what was then a very deprived part of the city. Her mother died from cancer when Sarah was just thirteen months old, and the children were brought up single-handedly by their father. Her twin sister developed meningitis in infancy and was left physically and intellectually impaired, requiring lifelong care from her father and siblings. But Sarah was academically able, leaving school with several Higher levels.

Colin is adamant that in the nearly three decades of their relationship there has been no hint of this jealous behaviour, nor of any other features to suggest mental illness or a personality disorder. Sarah's first episode started after a particularly difficult time: her twin had passed away from a heart attack, and one of her brothers had just taken his own life in the grounds of a mental health unit. The couple's initial thoughts were that stress had triggered this event. She had understandably become rather depressed, dealing with her grief when her first episode occurred. She had lost some two stone in weight. But the episodes continued even when she made a good recovery from her depression.

Despite the lack of a formal diagnosis made by psychiatrists during Sarah's admissions, Colin and Sarah report other symptoms that hint at psychiatric disease. It is not just pathological jealousy, a paranoia directed against her husband. During these episodes, her heightened suspicion extends beyond her immediate family. She believes that everyone is plotting against her. After one episode, a family member reported seeing her in the street. 'My wee nephew said: "You were hitting a man in the street." It came flooding back to me. I grabbed a man and started punching him. I grabbed his car keys and flung them away. The man got out [of his car], then punched me, and then I attacked him back and ran away. It was terrible. I [later] tried to apologise to the man, but a neighbour told me that he didn't want to see me, that he accepted my apology. That everyone was worried about me. I just wanted to make amends.' She does not recall why she attacked the man, but Sarah thinks he was just getting out of his car at the wrong place at the wrong time.

Even beyond this paranoia, and these delusions of infidelity and persecution, there have been delusions of another type. 'There were also some positive [delusions]. One time she thought Prince, the singer, was coming to our house for a party,' says Colin.

All these features point to Sarah's morbid jealousy being part of a broader psychotic illness, like schizophrenia – a more widespread deterioration in her mental health. But her psychiatrists remain uncertain, and there may be a little more to it than that.

* * *

On rare occasions, the sort of pathological jealousy seen in Sarah may even reach the attention of the courts. According to his defence counsel, Robert Mercati was a 'courteous and jolly raconteur with very human weaknesses – mainly, of course, his ever-increasing dependency on alcohol'. The barrister was at that time defending Mercati from a charge of theft.[17] He had been charged with a diamond heist in May 2003, from a display cabinet at the Berkeley Hotel in Knightsbridge, London. At that time fifty-four years old, he and an accomplice had acquired a duplicate set of keys and had made off with gems worth over £200,000 ($255,000). He had only been caught after a police appeal on the BBC's *Crimewatch* programme and was convicted and imprisoned for eighteen months. The gems were never recovered.

In early 2012, Mercati consulted a psychiatrist. He had begun to be totally preoccupied by thoughts that his wife, Margaret, was being unfaithful. He had hidden electronic bugs in their flat to spy on her, checking them furiously for evidence of infidelity, and had accused one of his sons of colluding with her. Mercati was prescribed anti-psychotic medication and over the months, until August 2012, his obsession reportedly improved. He had removed the bugging devices, and had become far less fixated, recognising that his jealousy had no basis in fact.

Unfortunately for Mercati and his wife, his 'very human weaknesses' had not disappeared though, and in late 2012, he was convicted of another theft, and spent a further few months in prison. Despite his medications being made available to him, he stopped taking them, and by the time of his release in January 2013, his delusions had worsened again. He was described as becoming 'enraged' by his wife's apparent infidelity again.

A few months later, on a May morning, Mercati had tried to force his way into their bedroom, shouting that his wife was in there with another man. Margaret had phoned for an appointment for her husband with a psychiatrist, and had also called her son, who raised the alarm. But within two hours of the phone call, Margaret was found strangled in the living room of their Bloomsbury flat, and Robert had hanged himself. The coroner's opinion was that this murder–suicide had been due to delusional jealousy.[18]

* * *

Mercati's case is certainly not the first or last example of morbid or pathological jealousy resulting in murder, either in real life or in fiction. It has been dubbed 'Othello syndrome', after the Shakespearean character who is convinced by Iago that his wife, Desdemona, is being unfaithful with Cassio, ending in her murder: 'But jealous souls will not be answered so. They are not ever jealous for the cause, but jealous for they're jealous. It is a monster begot upon itself, born on itself.'[19]

Jealousy is of course a normal emotion, and while it can result from many different kinds of social comparators, we usually consider it in the setting of intimate relationships. In this context, it has a clear evolutionary advantage, ensuring sole possession of a partner to allow propagation of one's own genes.[20]

Some scientists argue that sexual jealousy is a societal construct, not innate to us. Proponents of this view argue that some cultures, like the Inuit, share partners, but this ignores the fact that even in these societies, a common cause for spousal violence is male sexual jealousy. Several studies have shown that men experience greater levels of jealousy in response to the sexual aspects of an infidelity, while women show more jealousy regarding the emotional aspects of it. This finding has led some researchers to argue that jealousy has evolved for women in response to a threat to paternal investment, while for men it is about a threat to paternal opportunities.[21]

It is when jealousy has no rational basis, where the preoccupation is based upon unfounded evidence, where it results in extreme or unacceptable behaviour, that it becomes morbid jealousy. In fact, morbid jealousy may even be present when infidelity is actually taking place, if that belief is founded on insubstantial or non-existent evidence.

Even in the absence of murder, morbid jealousy can be extremely dangerous, and its presence makes doctors very nervous. The jealous partner will often become coercive and controlling, will repeatedly accuse and interrogate, check phone records and correspondence, pay surprise visits, or may monitor or limit their partner's movements or forbid the wearing of 'revealing' clothing. Repeated examination of bed linen, underwear, or indeed their partner's genitalia, may occur. Over half of morbidly jealous individuals have assaulted their partners, and many make threats to kill.[22] Violence can also be directed towards the imagined love rival, with cases of murder of the 'paramour' rather than the 'unfaithful' partner described. There are risks to the sufferer too – of depression, substance abuse, suicide – and risks to children, who may witness arguments or physical violence.

Morbid jealousy comes in different forms

Not all cases of morbid jealousy are the same, though. The roots of morbid jealousy may be very different, as in these two cases from a 1965 paper.[23]

The first is of a thirty-one-year-old machinist, whose wife was referred to a psychiatrist for excessive drinking. The wife reported that she would drink every night alongside her husband, but that the major issue for her was his 'insane' jealousy of her. He would check how long she had been shopping for, or even stop her from shopping at all. When she complained of menstrual cramps, he would accuse her of taking medication to 'get rid of someone else's kids'. At night, getting out of bed to use the toilet would result in accusations that she was going to meet another man in the darkness. On one occasion, while in the bathroom, a cat had made a noise outside, and he had jumped out of bed, chasing down her imaginary lover outside. He would call her ten times a day, and if she did not answer immediately, a torrent of accusations would ensue. This had all taken its toll on her: she was now a nervous wreck, and, she told the psychiatrist, if it were not for her children she would consider suicide.

Eventually, the husband was persuaded to see the psychiatrist too. He was described as 'a shy, quiet, obese man who seldom volunteered any information . . . He sat rigidly in his chair, seemed to lack a sense of humour and was a nail-biter.* He had little social life, occupied himself with studying the stock market and collecting stamps and coins.' Mooney, his psychiatrist, detailed his life story. In childhood, after the birth of a younger sibling, the man had temporarily lost control over his right arm, which had been attributed to 'nervousness'. He had been a shy and anxious youngster, who had ended up marrying a young woman from the neighbourhood after a two-year courtship, only a year after leaving school.

He confirmed what his wife had said about his behaviour. He told Mooney that, about a decade earlier, he had learnt that a friend of his wife had been unfaithful to her husband. He began to worry that his wife might do something similar. About two and a half years prior, his wife had gone out to buy something, and he had been struck by how long she had been away. This was the trigger for his increasing suspicion

---

* Historically, nail-biting was considered a 'neuropathic trait', a marker of neurosis, like sleepwalking, fear of the dark and bed-wetting. As a nail-biter myself, I am reassured by a paper that does not bear out this view.[24]

of infidelity. He said that their relationship was relatively happy: 'But they were incompatible sexually. He wanted intercourse every night, but he knew his wife did not enjoy sexual relations.' This was confirmed somewhat by his wife separately: 'Her husband was the only man she had ever been interested in . . . Although she did not get complete sexual satisfaction in intercourse, she was always co-operative. Recently, he had been making increasing sexual demands on her.' He was drinking heavily, about a pint of whisky every night, and had developed insomnia and irritability.

He was started on medication, and his behaviour improved somewhat, but he eventually stopped the drug as he thought it was making him sleepy. His jealousy again worsened, and he began to sleep with a gun under his pillow, planning to kill his love rival. On one occasion, he menaced his wife with a knife, and on another, after a party, became enraged, telling his wife: 'If she were frigid with him, someone else must be satisfying her.' She threatened to leave him and went to bed, only for him to wake her and hint at suicide. Having made similar threats before, and therefore not taking him seriously, his wife had simply said: 'Go ahead.' The machinist went into his bedroom, locking the door behind him, and fatally shot himself in the head.

The second case is of a forty-one-year-old woman, an aircraft worker and housewife. In the previous few months, her husband had changed jobs and had begun to share a ride to work with a woman who worked at the same company. Some marks on his clothes and on the car's upholstery made her suspicious, and she began to suspect an affair. She thought that objects were disappearing around the house, and that her husband was removing them to gaslight her. She started to chain-smoke, and became afraid to eat, as she believed her husband might be poisoning her food. She described her cigarettes as tasting different, and thought this was also due to poisoning. A wound on her leg, related to a varicose vein operation, took on new meaning: her husband was injecting her every night in this wound.

The husband, some thirteen years her junior, was her third. Her first marriage had lasted fifteen years, but her husband's binge-drinking and abusive behaviours had eventually led to divorce. The second marriage was on the rebound, and lasted a mere three months. The current marriage, however, had been generally happy for the last six years. Her husband denied any affair, and stopped sharing a lift into work with his female colleague.

The woman was admitted into a psychiatric hospital in very poor condition. She was depressed and anxious, rambling of speech and underweight, having been too scared to eat. She refused to wear her pyjamas as she feared her husband had poured acid over them. Within a few days of being on an anti-psychotic drug, she became less agitated, and her rigid beliefs regarding her husband's infidelity and efforts to poison her gradually settled. She was released from hospital after sixteen days and remained well.

Overtly, these two cases have obvious similarities. The intense jealousy, the profound conviction of an affair in the absence of clear evidence. The harm to the relationship, hurt to the partner, and damage to the sufferer, culminating in suicide in the first case. But in other respects, they are very different. In the first case, Mooney describes a man who had always been seen as 'odd': socially isolated, shy, humourless and with a history of 'nervous paralysis' in childhood. Apart from his pathological jealousy, his thought processes and his wider beliefs were untainted, except in the presence of a pint of whisky per night. In contrast, the woman had had a relatively 'normal' childhood and was not unusual in her personality or character before her illness.

From a psychiatric perspective, these two cases starkly illustrate that morbid jealousy can come in different forms. The woman in the second case shows clear evidence of delusions – unwavering beliefs firmly held in the absence of evidence to the contrary, and a hallmark of psychotic experience. These delusions encompass not only her partner's infidelity, but also her convictions surrounding poisoning or clothes soaked in acid. She is psychotic, and the delusions of infidelity are part of that psychosis. These sorts of delusions may sometimes be the first presentation of conditions such as schizophrenia, or may appear in the context of broader psychosis, as in her case. Occasionally, delusions of jealousy may be 'pure', in the absence of any other features of mental illness. Unlike the often rambling, bizarre and inconsistent nature of delusions seen in schizophrenia, people with these pure forms will often describe their delusions eloquently and apparently plausibly.

In the first of the two cases above, however, Mooney concluded that the pathological jealousy the machinist exhibited was not delusional. Instead, Mooney thought that it had an obsessional nature, due to the young man's background of chronic alcoholism and 'odd' personality. Obsessional jealousy is characterised by repetitive, intrusive and irrational thoughts, followed by compulsive rituals of checking or seeking reassurance from the partner. Sufferers will recognise that their jealousy is without foundation. They are ashamed of their thoughts and know that these intrusive thoughts are against their conscious wishes, causing terrible distress. While obsessional jealousy can be considered as 'lesser', the consequences of it can be just as terrible, as the young machinist showed. In practice, there does appear to be an overlap between the two types, and sometimes a distinction can be difficult.

For Sarah (as with the latter case of the woman with psychosis), the origins of her pathological jealousy do not simply seem a function of an abnormal personality or past experiences. She too has unwavering beliefs surrounding infidelity, in the presence of evidence to the contrary. Her delusions, while centred on aspects of jealousy, are broader – encompassing Prince coming to a party, for example. She is overtly suffering from brief psychotic episodes. Their brevity, and that she is entirely mentally stable between them, are their redeeming features for Colin.

Psychological and neurological origins of jealousy

The underlying triggers of morbid jealousy are obscure. There is a view that people (usually men) who have attachment issues, especially who are anxious or fearful, may become increasingly agitated and preoccupied about their partners' attachment to them. These sorts of attachment issues correlate strongly with particular features of personality: feelings of unworthiness, a negative view of oneself, anxiety about rejection and abandonment, instability of mood or anger, and difficulty inhibiting impulses. Some of these personality traits may predispose to people distorting and incorrectly interpreting events, leading them to make false assumptions and initiating the path to morbid jealousy.

Other proposed factors include feelings of sexual dysfunction (here Freudian views of real or imaginary perceptions of having a small penis arise[25]) and associated feelings of inadequacy; cultural factors (such as ownership of one's partner); and alcohol and substance misuse. Alcohol abuse, in particular, is a very common factor in morbid jealousy, as in Robert Mercati's case, and amphetamines and cocaine have both been described as triggering delusions of infidelity, which may persist despite stopping their use. This association with substance abuse may be explained by alcohol or drugs freeing normally inhibited thoughts or behaviours, but it may also be that people with personality traits that predispose to morbid jealousy are turning to these substances to try to cope.

Somewhat confusingly, none of this seems applicable to Sarah. Not only does she show no signs of mental illness outside these brief, discrete episodes of pathological jealousy mixed with other delusions, but there are no long-standing personality traits, no alcohol or substance misuse that might explain her condition.

* * *

Thus far, morbid jealousy has been very much framed as a psychological or psychiatric state or condition. But as I have hinted at, my view is that the psychological and the neurobiological are indivisible: that all emanates from the brain. Perhaps we are simply too ignorant – too tethered to existing scientific techniques and research methodologies – to understand the neurobiological basis of our psychologies, but ultimately we will find the answers (I hope).

If that is the case, then what evidence do we have that jealousy as a trait has an underlying neurobiological basis? One study asked participants to read scenarios, imagining themselves to be in romantic situations: some happy, others threatened by a romantic rival. During these imagined scenarios, they were asked to rate their happiness and jealousy levels, and were put into a functional MRI scanner. The results revealed that romantic jealousy produced activation predominantly in deep centres of the brain,* while romantic happiness activated rather different areas.[26]

While this study confirms neural correlates of jealousy, it is

---

* The basal ganglia, thalamus and middle cingulate cortex.

difficult to assess if those visible changes on functional MRI are the underlying cause of jealousy or reflect the experience of it. However, individuals with neurological disorders provide more definitive proof that brain changes can result in jealousy. One such case involved a forty-year-old school teacher, previously entirely well. She had been brought in to see a psychiatrist due to increasing suspiciousness for the last few months, and strongly believed that her husband was trying to poison her and was having sex with a number of younger women. Previously relatively shy, she had become increasingly aggressive and violent towards him. Initially treated with antipsychotics to limited effect, a brain scan was done to investigate, and a large benign tumour identified. Its removal caused a dramatic improvement in her delusional jealousy.[27]

This school teacher is one of many cases in the literature of people with neurological diseases who have developed morbid jealousy. It has been estimated that some 15 per cent of people with morbid jealousy have an underlying neurological cause. The range of conditions is surprisingly wide: traumatic brain injury, degenerative conditions like Parkinson's or Alzheimer's, strokes or tumours. Indeed, delusional jealousy seems more common in psychosis that is due to neurological causes rather than conditions like schizophrenia or alcohol-induced psychosis – one study found that these types of delusions are found in 7 per cent of people with psychosis of neurological origin, 2.5 per cent of people with schizophrenia and only 0.1 per cent of people with depression.[28] Efforts to home in on a particular location in the brain, a 'jealousy centre', have been fruitless, which is perhaps unsurprising given the various factors implicated in such a complex and ancient emotion.

This relationship between morbid jealousy and brain disease has been known for a very long time. Indeed, it is described in

Alois Alzheimer's very first publication detailing his eponymous syndrome. In his landmark paper of 1907, titled: 'Über eine eigenartige Erkrankung der Hirnrinde' ('On an unusual illness of the cerebral cortex'), Alzheimer relates the case of a fifty-one-year-old woman from the 'insane asylum' of Frankfurt-am-Main: 'The first symptom . . . was the idea that she was jealous of her husband. Soon she developed a rapid loss of memory. She was disorientated in her home, carried things from one place to another and hid them, sometimes she thought somebody was trying to kill her and started to cry loudly.'[29] Alzheimer went on to describe the very typical memory disturbance of Alzheimer's disease, but it is striking that the first symptom of the first patient ever formally described in the literature was that of morbid jealousy.

It is not just Alzheimer's disease that is associated with morbid jealousy, but other forms of dementia too.[30] Patients with a very different type, dementia with Lewy bodies,* which has commonalities with Parkinson's disease, are even more likely to exhibit morbid jealousy, and sometimes report visual hallucinations that confirm their jealousy – visions of their spouse in a sexual situation, of the partner having an affair in the house or images of a child their partner has had with their lover. This all suggests that neurological decline in combination with psychosocial factors potently stokes the fire of pathological jealousy.

Curiously, in these patients the risk or severity of these delusions is not associated with the degree of cognitive impairment. One important factor that drives this phenomenon is low self-esteem and feelings of inferiority, and it has been suggested that while mild cognitive decline may give rise to feelings of inferiority,

---

* Dementia with Lewy bodies often results in visual hallucinations. It is possible that patients misinterpret these hallucinations as evidence of infidelity, or perhaps their underlying preoccupations of infidelity are integrated into their visual hallucinations, modifying their content or interpretation.

in the later stages of dementia the sufferer no longer has this insight. Without feelings of inferiority, there is no envy, no jealousy.

Indeed, for Sarah too there are some indications that the origins of her mental disturbance may be related to a neurological disorder rather than psychiatric disease. One of the notable aspects of her case is the relatively sudden onset of her jealousy and paranoia, and its equally sudden termination. Her periods of psychosis last only a few days, resolving spontaneously without a definitive treatment. When she was forcibly detained in a mental health unit, she was back to her old self within a couple of days. 'She was released after eight days, because there was such a dramatic turnaround, and she wasn't on any medication. Totally back to normal, so they let her go,' says Colin. This aspect alone, the brief nature of her episodes with rapid normalisation would, in itself, be rather unusual in psychiatric conditions.

Then there is the pattern of memory that she describes. When she comes out of her episodes, she does not immediately have recollection of events. She reports these flashbacks, snippets of memories returning, something that again would also be unusual in psychosis of a psychiatric cause.

On two occasions Sarah has been found to have urinary tract infections during these episodes, having suffered from them since childhood. Colin and Sarah wonder if these infections might be poisoning her system, causing her to lose mental functioning, to loosen her grip on reality. It is true that infection can cause acute confusion, sometimes with sudden onset, termed delirium. So too can medications or altered levels of electrolytes in the bloodstream: changes outside the brain – be these inflammation, drugs or other factors – causing brain dysfunction from a distance. It is more common in the elderly, in people whose brains are not working quite as well in the first place, but can affect the young too. However, neither Sarah nor Colin report any of the other typical features of delirium – drowsiness, inattention or disorganised thinking, for example. People can suffer delusions in delirium, but these are usually rather poorly formed, in contrast to Sarah's more specific and complex beliefs.

There is another possible explanation, however. There are three potentially highly relevant features to what Sarah and Colin tell me. The first is that Colin can often tell when these episodes are going to happen. He describes a phase in the day or so beforehand when she becomes quiet, subdued, with limited interaction, before she then becomes clearly psychotic. The second is an association with her menstrual cycles. All three episodes have occurred within the prelude to her periods and have abated when her periods have come.

The final feature is one that really cannot be ignored. Sarah has been given a diagnosis of non-epileptic seizures, fits that do not have a basis in abnormal electrical activity in the brain, previously considered 'psychosomatic'. For the last thirteen years, she has had the occasional convulsion, about once a year. Crucially, however, although she has had MRIs of her brain that show it to be structurally normal, she has never had an EEG to study the electrical activity of the brain. And, talking to Colin, the description of her seizures makes me wonder if they do actually have a basis in epilepsy. For a start, they often arise from sleep, which is less common in non-epileptic seizures, although not totally impossible. He also describes the very characteristic features of epileptic seizures: the rolled-back open eyes, the clenched jaw, the typical rigidity and rapid shaking; the excessive salivation, urinary inconti-nence, the noisy breathing; sleepiness and confusion after the event. Each of these features in isolation does not discount non-epileptic seizures, but in their totality would certainly make me want more definitive proof that they are not truly epileptic. And epileptic seizures can be triggered by infection, and in some women are more likely to occur in the prelude to their periods. Indeed, during one of her admissions, Sarah had seizures while an inpatient, and improved rapidly afterwards.

So, if Sarah's seizure diagnosis is incorrect, that she actually has epilepsy rather than non-epileptic seizures, might this be an explanation for her psychotic episodes – like those patients in Chapter 1 with psychosis after their seizures? It is possible that she is having small seizures – electrical seizures without more obvious manifestations of epilepsy, apart from the subtle changes in her behaviour Colin notices as a prodrome to her more dramatic events. Triggered by infection or her menstrual cycle or both, clusters of small seizures are disrupting normal brain function, giving rise to transient psychotic symptoms.

I readily admit that this possibility is conjecture, and based upon Sarah's clinical history as told to me, but thus far there has been no other definitive diagnosis for her. There is no strong evidence for mental illness, brain tumour or anything else. And she would not be the first person with pathological jealousy as a manifestation of epilepsy.[31] For the moment, assessments are ongoing, and hopefully time and other investigations will tell. If my supposition is correct, however, then it provides a clear course of action to rid her of these episodes of toxic jealousy: treatment with anti-epileptic drugs.

* * *

At the core of envy and jealousy are the primitive experiences of want, of need, anger and resentment. Primaeval emotions fundamental to survival and competition. That these so-called sins, like the others, originate in the metaphorical and literal deeper recesses of the brain should be of no surprise. These brain regions most strongly linked to emotion and to biological drivers such as pleasure, hunger and thirst are ancient. In evolutionary terms, they are among the oldest regions of the brains of vertebrate animals – the 'reptilian complex' and 'paleomammalian complex', proposed to be derived from reptiles and early mammals respectively, of MacLean's model of brain evolution and behaviour.* In humans, and other primates, these 'sins' are balanced by an ability to inhibit them, to suppress them through our ability to reason, our propensity for rational thought.

And, as with the other sins, it is when these emotions are unchecked that a useful trait risks a crescendo, amplified to levels that no longer benefit us, but instead jeopardise our health and that of those around us, and even life itself. When disorders of the brain disrupt the delicate equipoise between our basic instincts and our virtuous nature, those normal emotions, fundamental to humanity, to our advancement individually or collectively, run amok.

---

* Paul D. MacLean first proposed the 'triune model' of the evolution of the brain in the 1960s. He divided the brain into the reptilian brain (the basal ganglia), the paleomammalian brain (the limbic system) and the neomammalian brain (the cerebral neocortex), although similar views of the brain are much more ancient in Western thought. This model has been much criticised and disputed,[32] and is now widely considered overly simplistic or frankly incorrect. Those primitive aspects of our brains do not originate in reptilian ancestors. In more general evolutionary terms, however, it is clear that some aspects of brain function and structure evolved much earlier than others.

5.

Sloth

Thou seest how sloth wastes the sluggish body, as water is corrupted unless it moves.

Ovid, *Ex Ponto*, Poem 5

Rarely, I see a patient in my clinic who illustrates the true meaning of sloth – the proper nature of this 'sin'.

Such an individual is a young man I have written of before, so remarkable is his story.[1] I first met him with his wife, their five-year-old child left with family in their hometown. Since the birth of this child, he had been struck down by an overriding need to sleep, for twenty hours a day. His mysterious condition had rendered him not only helpless in the care of his child, but had left his wife to look after him as well as the baby. He had missed out on his baby growing, walking and talking, indeed all aspects of family life, over five years. Multiple medical investigations and evaluations had not provided a diagnosis, treatment or progress. He remained largely bedbound, while his family continued life around him. However, a prolonged admission into the sleep laboratory quickly demonstrated that, while the video showed that he did spend twenty hours a day in bed apparently asleep, his EEG revealed something very different. For about twelve of those twenty hours, his brainwaves clearly showed him to be fully awake, feigning sleep.

After a rather delicate and awkward consultation explaining these results, I gently suggested that his 'sleeping' issue might have a psychological basis (I did not admit to my suspicion that he was probably malingering – pretending to be sleepy in an effort to escape his parental duties), and we agreed that he should be admitted to a neuropsychiatric ward. Some three days prior to his planned admission, I received a phone call from his wife, telling me of a miracle. That he had suddenly woken up, that his problem had been cured. The joy in her voice radiated from the telephone, as she described him being up and around, helping look after the child, doing the shopping, the cooking and the cleaning. He did not attend the neuropsychiatric ward, and I never heard from them again. His wondrous recovery under threat of further investigation or treatment reinforced my view that he had been caught out.

To my secular mind, sloth is the oddest inclusion in the list of Seven Deadly Sins. After all, a little laziness is hardly as harmful as wrath, envy or lust. And depending upon your interpretation of what exactly sloth is, it can be a very positive trait. For some, sloth represents excessive sleep or sleepiness. From my own professional perspective, many of the patients I see simply do not sleep enough, suffering from a dearth of 'sloth', and reaping the repercussions: the fatigue, cognitive difficulties, mood disturbance and general inability to function. A little more sloth would definitely be a good thing. At the other extreme are those patients who sleep too much, through neurological disorders such as narcolepsy and idiopathic hypersomnia. People for whom sleep seizes them, or simply does not refresh them. When awake, however, they are not lazy. They desire nothing more than to engage with the world, to achieve, to enjoy, to participate.

Other forms of sloth also clearly have benefits. When we are ill, we feel tired, lack motivation and avoid activities and people – changes in our actions termed sickness behaviours.[2] It would be reasonable to assume that these effects are due to toxicity, down to the bacteria or viruses invading our bodies. However, rather than a diffuse result of infection, these sickness behaviours appear to be carefully orchestrated by our brains, adaptations to redirect energy and resources, to reorganise physiological functions, all in an effort to fight our bodily invaders. This behavioural response to infection appears to be the duty of a specific set of neurones within the brain.*

Various dictionaries, however, define sloth rather more precisely. These definitions centre on inactivity, akin to these sickness behaviours rather than sleep, on a disinclination to act, to exert oneself or to work. A lack of effort, an idleness or indolence, an inability to generate action. Sloth is the lack of motivation. As with the young man above, whose 'sleep disorder' was a convenient excuse for his inaction, or the biological effects of a good dose of flu.

In the religious context, it is this lack or failure of motivation that constitutes what has been described by some theologians as the gravest of sins. In the Christian tradition, the description of sloth as a cardinal sin originated in the monastic communities of the Egyptian desert in the fourth and fifth centuries of the Common Era. It represented the boredom, fatigue and listlessness of lunchtime in these communities, of monks in the afternoon lull plagued by thoughts of fleeing their religious life and their commitment to God.[4] Later, as Saint Thomas Aquinas taught, sloth became synonymous with a lack of will to do

---

* In mice, a well-defined group of brain cells become more active in response to chemical signals of infection. If you stimulate these cells, the mice develop sickness behaviours – reduced food and water intake, and reduced movements. Block the activity of these brain cells, and you lessen the effects of infection signals on the behaviour of the mice. The 'sickness' neurones are located in two specific areas of the brainstem called the nucleus tractus solitarius and area postrema.[3]

God's work, an apathy towards doing good, to contribute to the spiritual and physical well-being of oneself and one's community.

Medical 'sloth'

Beyond this religious perspective, however, from a medical viewpoint, sloth by another name is very commonly seen. The lack of motivation that is the essence of sloth is described in a high proportion of individuals with a wide range of neurological and psychiatric disorders. In the neurological world, this syndrome is usually referred to as apathy* – diminished motivation to engage in physical, cognitive or emotional activity – and is seen in many common neurological disorders such as stroke, Alzheimer's disease and Parkinson's, as well as other rarer conditions. In the psychiatric world, this state of being, almost a hallmark of major depression, and seen in other conditions like schizophrenia, is more typically referred to as anhedonia, the loss of interest in previously rewarding activities that produce pleasure.[5] It is this lack of pleasure, the absence of reward, that is considered the root cause of lack of motivation in anhedonia. The distinction is subtle but definite. In anhedonia, the lack of an emotional driver, of the gratification associated with an action, is the principal determinant of this failure of purpose.

At their core, these various phenomena imply that something in the brain has gone awry, that the way reward is processed to motivate behaviour within the brain or mind has been disrupted

---

* Apathy, derived from the Greek *apatheia* ('without passion'), has not always had negative connotations. For the Stoics of Ancient Greece, followers of the school of philosophy founded by Zeno of Citium in about 300 BCE, apathy was a state of being to be pursued. To be insensitive to the *pathē*, the emotions and passions of pain, fear, desire and pleasure, was to achieve a calm and quiet soul in the face of human suffering.

or altered. The absence or diminishment of reward obtained through action, social contact or other pleasurable events, such as eating or sex, results in a deep-rooted lack of motivation, a disinclination to do anything. And regardless of the underlying cause, increasingly evidence points to these phenomena having a shared neurobiological basis.

The components of motivation

'Sloth' may be conscious, as in the young man apparently sleeping his way through life, or unconscious, as in those individuals with apathy or anhedonia in the context of neurological or psychiatric disorders. To understand sloth in all its forms, however, requires some basic knowledge of the constituents that motivate us all.

Neuroscientists have long wrestled with the underlying psychological or behavioural changes that give rise to apathy. Motivation ultimately boils down to the making of decisions to expend effort to seek reward, and this process may go wrong at multiple stages.

Consider waking up one morning and seeing the sun shining outside your bedroom window. You may decide to go for a walk, find a nice spot in the park, take a blanket and a book, to read in the sunshine. Or you may go to the shops, buy a bottle of wine and some food for a picnic, and pick up the phone, arranging to meet some friends. These decisions, the exertion of effort with the reward of sunshine, a glass of wine, a good lunch or social interaction with friends, have a number of different components to them.

The first component is that of option generation – the ability to produce different possibilities for behaviour; in this setting, to go outside, to phone a friend or to continue lying in bed. If you cannot even think of the various options open to you, then this might be the first hurdle in getting up and doing something constructive with your day. Before you can even make a choice to act, you need to think of what choices are available.

In one study, patients with schizophrenia were assessed for their degree of apathy and presented with twenty short real-world scenarios.[6] They were asked to generate options for what they might be able to do in these scenarios. The degree of apathy was directly correlated to a poorer ability to generate options. Similar findings have been found in healthy people: the more apathy you have, the worse you are at generating options.

Even once you have thought of these various options, you need to be able to select a course of action, to choose between those options. Do you pick the book or the bottle of wine, the peace and solitude of a quiet corner of the park or a raucous gathering of friends? The decision you make is dependent on several factors. It might be more fun to call some friends and have a party, but it requires more effort, and may take longer to organise. However, if the weather turns, you may be able to reconvene in a restaurant or a pub. If you go for the sunbathing option, you will simply have to pack up and go back home. Which option to take requires an evaluation of multiple elements: the reward likely to be obtained by any option, how much effort is required related to the reward anticipated, the likelihood of obtaining the reward, and how much time you will have to wait for it. It is that balance between effort and perceived reward that guides your decision. You may calculate that the pleasure you might derive from meeting your friends outweighs the risk of your party being rained off. If you have seen your friends earlier in the week, you may decide that you will derive more pleasure from some time on your own.

Within this aspect of motivation, anticipation of reward is

fundamental. We are spurred to act by the dangle of the carrot in front of us. Motivation to achieve a reward is driven by anticipation, and this can be measured through changes in our physiology, such as heart rate increase or dilation of our pupils. When faced with a potential reward, our pupils dilate, and the degree of that dilation is proportionate to the value of that reward. One study utilised pupillary size to measure anticipation of reward in normal individuals.[7] Subjects were asked to look directly at a disc at the centre of a screen in front of them, and a voice would tell them the maximum reward available for that particular test, ranging from nothing to fifty pence. After a brief pause, the central disc would disappear, and a new target disc would be displayed randomly on the left or right. The proportion of the maximum reward obtained would relate to the speed at which they moved their eyes to fixate on the new disc. The researchers used an infra-red eye tracker camera to monitor pupillary diameter, and speed of eye movements. What they found was that pupillary dilation was significantly increased by the amount on offer, i.e. the value of the reward, although this was less marked in older, healthy subjects than younger ones. The same study was performed in individuals with Parkinson's disease. Those with high levels of apathy showed blunted pupillary responses compared to those without apathy, suggesting that apathy is associated with relative insensitivity to reward, at least in those in Parkinson's disease.*

In addition to evaluating how much you want that glass of wine or that book, you also need to decide how much effort you are willing to put in to obtain either of these rewards. How

---

* When these individuals were exposed to medications boosting dopamine levels, they demonstrated an increased sensitivity to reward, as measured by their pupillary responses. This increased response to potential reward may go some way to explaining some of the behavioural changes seen with these medications, described with regard to lust, previously.

far from your nose that carrot is being dangled. Effort may be physical, as in how far you need to walk to go to the super-market to buy your food for the picnic, or mental, as in how much organisation and phone calls are required to gather your friends. The amount of effort you expend in the pursuit of a goal or reward is likely to be a function of the odds of achieving that goal, not only the anticipation of the potential reward. If the likelihood of a reward is low, you are much less inclined to go for it in the face of much effort.[8] The assessment of effort, however, requires a yardstick, a metric of energy expended, and how you measure effort may be crucial to motivation.

And of course, the effort that you go to is going to be influenced by the pleasure that that reward brings you, once you get it – the size of the carrot. If reading brings you little joy, then you are much less likely to pursue the book option.

So apathy – 'sloth' – may be related to a breakdown in any of these various aspects of decision-making to achieve a goal: the generation of options, the ability to weigh up the effort of these options and their likely reward, and the degree to which you want or like that reward. What then do we know about where in the brain these processes might be happening?

A good place to start is by looking at those neurological disorders in which apathy is a common feature, and which regions of the brain are associated with those conditions. One such disorder is frontotemporal dementia (FTD), a form of dementia much rarer than Alzheimer's disease, which tends to affect younger individuals. FTD encompasses patterns of neuro-degenerative disease that have a predilection for behaviour, language and the mental processes (termed executive function) that allow us to plan, maintain focus, juggle multiple tasks and remember instructions. There is sometimes significant overlap in how patients with Alzheimer's and FTD first appear, particularly early on in the disease.* Even now, with huge advances in our understanding of the genetic, pathological and imaging changes that differentiate the two types of dementia, establishing a firm diagnosis rapidly can be difficult: for more than 40 per cent of people with FTD it can take over a year to have the diagnosis established, seeing multiple doctors on multiple occasions.⁹

FTD, as its name suggests, has a predilection for the frontal and temporal lobes, and the way it manifests in people relates directly to the functions of these areas of the brain. There are different subtypes of FTD,† but there is one variant that has a distinct relevance to apathy. This subtype is characterised by predominant degeneration of the frontal lobes. As would be expected, given the role of the frontal lobes in our decision-making and regulation of our behaviour, patients with this variant, termed behavioural-variant FTD (bvFTD), exhibit

---

* The distinction between the two types of dementia is increasingly important. A firm diagnosis informs the patient and their family as to what to expect, but as our understanding of the genetics of these conditions and the underlying causative pathological process progresses, it also influences targeted treatment. Alzheimer's disease more typically affects the temporal lobes, especially in deeper parts involved in memory such as the hippocampus, or the parietal and occipital lobes at the back of the brain, which are involved in visual and spatial processing. With the former, patients will exhibit prominent memory loss, particular for recent memories, while in the latter, the degeneration of the parietal and occipital lobes result in problems with reading, recognition of faces, mental arithmetic and visuospatial disorientation.

† Where the degenerative changes predominate in the temporal lobe, people will often lose knowledge of the meanings of words, as well as the words themselves. Someone with a stroke or Alzheimer's disease, when looking at a picture of an elephant, may lose the word 'elephant', but will know that it is an animal with a trunk, that lives in Asia or Africa, that has a wrinkly grey hide. Someone with this variant of FTD will also be unable to say that the picture is an elephant, but will also lose the knowledge associated with this category of animal. It is not purely a speech issue, but a meaning issue too. The speech itself may be fluent, but the content of that speech will gradually become meaningless. Another variant is much more focussed on the construction of spoken language. These individuals will usually not lose words, at least not early in the disease process, but they will make errors in grammar or sentence structure.

unusual changes in personality and conduct. They will become socially disinhibited, will lose sympathy or empathy, develop ritualistic or compulsive behaviour, or will show changes in their dietary preferences or even become hyper-oral (compulsively put edible or even inedible objects in their mouths). Patients often exhibit childish or socially inappropriate behaviour, may eat the same food constantly, such as peanut butter and jam sandwiches daily, or show more subtle behavioural changes such as poor judgement or slight paranoia.[10]

However, another hallmark of bvFTD is a very prominent loss of interest, a lack of initiation of action, and a social or emotional withdrawal from those around them. This apathy is not something seen in these patients prior to their illness but emerges with its onset. Furthermore, it is not simply a function of the severity of disease and its consequences but seems to be a fundamental aspect of the changes occurring within the FTD brain, from the very early stages of the disease. Apathy is seen in over 90 per cent of people, even at very early stages. The brain scans of those with FTD and prominent apathy show increased loss of grey matter in several regions of the frontal lobes, importantly including the pre-frontal cortex.[11]

FTD is not the only condition affecting the frontal lobes that is associated with apathy. Strokes and brain trauma affecting this region are also strongly linked to this trait. Some of these brain regions are fundamental to the generation of options, already implicated in apathy from a psychological perspective. They are also important for planning, for the organisation of actions in order to achieve a goal. Other regions within the frontal lobes are lynchpins in the translation of emotions of mood into action or behaviour, and indeed damage to these areas results in emotional blunting. Areas like the vmPFC are strongly anatomically connected to the limbic system. Without an emotional component to guide us, it is difficult to properly evaluate the consequences of our choices and actions.[12]

All this points to one conclusion: that the frontal lobes are fundamental to our levels of motivation and the basis of apathy. They are crucial to many aspects of the processes that guide us to act, to move, to strive – the generation of options, weighing up cost versus benefit of energy expenditure, and ultimately the selection of a course of action.

But there is another group of conditions, also associated with apathy, that affects other regions of the brain. They reflect these multiple components of the choice to act.

* * *

Rhett is a man of many contradictions. He resembles the all-American guy, with sunswept face, pale beard and baseball cap, looking every inch the cattle rancher that he is. Yet he is the descendant of Lebanese Maronite immigrants to the United States, who arrived in the early 1900s with no English, and no money in their pockets. His grandfather embodied the American Dream, starting school speaking only Arabic before moving out West, buying swathes of land and dying a very rich man. Rhett now runs a stocker-feeder operation of a thousand head of cattle on one of the family's smaller plots in Kansas, although in total the family owns some tens of thousands of acres across Kansas, Oklahoma and New Mexico.

He is also a man of science. After going to college to study agriculture and returning to the ranch, a car accident on the highway outside his home ended in tragedy. One high-school boy was killed, another left paraplegic. 'Man, I need to get some training, so I can help somebody,' he recalls thinking, before volunteering as a paramedic in a rural ambulance service, and then going to medical school. He graduated as a physician assistant, and still works shifts at his local hospital.

He is at the same time a man of profound faith, describing himself as a follower of Jesus Christ, preferring this term to evangelical Christian, which he says has negative connotations for some people. Despite being brought up in a Catholic family, he only found religion in his early twenties. Until this point, he had lived a gilded life in a very wealthy family – land, cars, a private plane. He reports a life of hedonism, materialism, women and alcohol. 'I was known to be a player,' he says. 'On the outside, everyone thought I was a nice guy. I was nice to old ladies, helped them cross the street. But there was a side of me that was very at home in the bar in the dark, with whoever, whatever, wherever. I had money, I had alcohol, I had sex. I had a great career in the family business. I had all these things that the world says you are supposed to have to be happy.' Then he had a deep personal crisis; a realisation that despite having everything, he was empty inside, that his life had no meaning. 'So, I really started searching. I studied all the religions. And the only thing that made sense to me, when I looked at all of them, was Christianity. All the other religions are Man trying to reach God, trying to perform for God, trying to make themselves good enough for God. Whereas Christianity is Jesus Christ reaching down to Man; it was completely diametrically opposed. When you really study it, Christianity, true Christianity, not religion, is where people are still working to be good. True Christianity is God reaching us through Jesus Christ.'[13] Despite my utterly contrasting world view, as we chat, I warm to Rhett, finding him funny, thoughtful, occasionally irreverent and very open-minded.

It was in his 'player' years that Rhett first met Becky, his future wife. 'I had moved back from college, and I was still a party boy, just interested in girls and beer. We went for a couple of dates, but we did not hit it off at all.' He describes her as strikingly beautiful, 'mesmerising', a mix of Bayou Cajun from south Louisiana and the Choctaw tribe of Oklahoma. Even though their dates had been disastrous, and their very different backgrounds – she from a poor family, working as a hairdresser, he wealthy and privileged – he could not get her out of his head. After his embrace of religion – 'God straightened my life out,' Rhett says – 'I had to ask her out again. I was almost obsessed. So I finally convinced her to go out with me again. I knew I was going to marry her within two weeks. And I just pursued her until she married me. I'd never been so infatuated with a woman in all my life.' They married when Becky was twenty-one, and Rhett was twenty-five.

Rhett was aware of Becky's background before they were married. Her paternal grandmother had been one of seven siblings, in rural south Louisiana. Only two of the seven escaped a hereditary degenerative brain disorder called Huntington's disease, and unfortunately Becky's grandmother had not been one of them. Becky's father was born to a woman with Huntington's, and early on it had become apparent that he had not avoided it either. 'He had difficulty dealing with life,' says Rhett. 'He couldn't hold down a job. He was just odd.'

Rhett recalls his father-in-law working as a janitor at a high school when he was sacked due to beating up one of the students. 'He had outbursts of anger. He would repeatedly get fired from every job. But, on the other side, he was one of the kindest, most caring individuals you ever met. He didn't have a dime to his name, but he would give you what he had. He was just a really sweet guy. But then we would have these episodes of craziness.'

After each spike of bad behaviour, Becky's father would go to the doctor. 'He would say: "I know I have Huntington's. My mother has it, my sister has it. The majority of my cousins have it. All but two of my uncles and aunts have it." And they would say: "No you don't have Huntington's. You are just depressed, or bipolar." Because he didn't have any chorea.'

The chorea of Huntington's disease

George Huntington was only twenty-one when he graduated from Columbia University in 1871. He was working as a family doctor in East Hampton, close to the eastern tip of Long Island, New York, when he observed cases of dementia and abnormal movements in his middle-aged patients, strongly running in families.[14] His father and grandfather, also local doctors, had seen these patients before, all of whom were descended from one man, Jeffrey Francis, who had arrived there from England in 1634. Francis had brought with him to the New World the gene responsible for this neurological disorder, a terrible inheritance for his progeny.* Huntington, building on the work of his father and grandfather before him, published an essay in the *Medical and Surgical Reporter* of Philadelphia in April 1872, detailing his patients and this condition. The essay was titled 'On Chorea', and described the abnormal movements of these individuals, chorea being derived from the Ancient Greek for dance:

> The name 'chorea' is given to the disease on account of the dancing propensities of those who are affected by it, and it is a very appropriate designation. Its most marked and characteristic feature is a clonic spasm affecting the voluntary muscles . . . The disease commonly begins by slight twitchings in the muscles of the face, which gradually

---

* The example of Jeffrey Francis was later held aloft as justification for immigration controls by Charles B. Davenport, director of Cold Spring Harbor Biological Laboratory and founder of the Eugenics Record Office, in a paper of 1916.

increase in violence and variety. The eyelids are kept winking, the brows are corrugated, and then elevated, the nose is screwed first to the one side and then to the other, and the mouth is drawn in various directions, giving the patient the most ludicrous appearance imaginable . . . As the disease progresses the mind becomes more or less impaired, in many amounting to insanity, while in others mind and body gradually fail until death relieves them of their suffering. When either or both the parents have shown manifestations of the disease, one or more of the offspring invariably suffers from the condition . . . It never skips a generation to again manifest itself in another . . .[15]

As so frequently occurs in the history of medicine and eponymous conditions, others had described these patients in the medical literature several years earlier, but it was George Huntington's name that was immortalised through the disease. The chorea – these dancing-like movements that can affect face or limbs; the inability to keep still – was the hallmark of the condition, and its absence in Becky's father's case led to his diagnosis being delayed for many years.

Our genetic and clinical understanding of Huntington's disease has progressed since Dr Huntington's famous paper of 1872. Fortunately, we now better know the cause and its various manifestations.

Each of us carries two copies of the *huntingtin* gene, the protein product of which is produced in all our cells, but particularly in the brain. The huntingtin protein has an important role in preventing cell death and promoting the development of neurones.

The gene itself has a property that is commonly found in many other genes. Our genetic code comprises a sequence of

3 billion letters, each corresponding to a modified sugar mole-cule, chained together within the structure of DNA. These molecules are not identical, and there are four different molec-ular structures. These molecules, termed nucleotides, are called guanine, adenine, cytosine and thymine (the letters G, A, C and T respectively). Each triplet of letters acts as the template for a different amino acid, the building blocks of protein that are strung together in chains. It will never cease to be a wonder to me that our entire anatomy and physiology, our very existence, is encoded by four letters.

Scattered throughout our genome, there are areas of repetitive sequences of all sorts, misprints or a stuttering of the genetic code, but especially trinucleotide repeats – where the same three letters are duplicated over and over. Such a repeating region exists within the *huntingtin* gene as well. Within its sequence is a region of the same three letters, CAG, repeated multiple times. Each CAG encodes the amino acid glutamine in the final product of the huntingtin protein, and so these CAG repeats translate into a chain of glutamine within the structure of the protein.

The number of CAG repeats within the gene varies, but for healthy individuals there will be up to twenty-six CAG repeats. If the repeats are too numerous, however, if the glutamine chain in the huntingtin protein is too long, the structure and the function of the protein are altered. A long repeat sequence profoundly changes the protein's characteristics, leading to death of brain cells, and the degeneration of the brain that ends in disease. Carry forty or more CAG repeats in one of your *huntingtin* genes and you will definitely get Huntington's disease. With thirty-six to thirty-nine repeats, you may get it, but not necessarily – and if you do, it may start later and progress more slowly. If you carry a huge number of repeats of the three-letter code CAG within your *huntingtin* gene, sixty or more, then you may even get the disease in childhood.[16]

There is another unfortunate aspect to the genetic underpinnings of Huntington's disease. The production of sperm or eggs relies on the copying of genetic information, as does all cell division and replication. This process is usually highly reliable, since our health and survival are entirely dependent upon it. But as we have already seen with Alex and her Prader-Willi syndrome causing excessive eating (see Chapter 2), sometimes our bodies blunder. In Huntington's disease, the molecular machinery that reads and copies our genetic code can struggle with these repetitive sequences, and lose its place or slip – like reading the same word on a page repeated multiple times and losing count. Thus, if you have a long repeat sequence, there is a risk that, when your sperm or eggs are produced, further repeats can be inserted in the copying of the DNA sequence, resulting in an expansion of the repeat length.

This has two potential consequences. The first is that for individuals who carry between twenty-seven and thirty-five repeats, who will not develop the disease, an expansion may be transmitted to their offspring, resulting in the disease-causing threshold of thirty-six repeats being met. Secondly, in individuals who have Huntington's disease, the repeat size transmitted to their children may increase, resulting in more severe disease or a younger onset – a genetic phenomenon known as 'anticipation'. The production of sperm is inherently a less stable process, and so anticipation is particularly seen in fathers with Huntington's disease, rather than mothers.[17]

In addition to insights into its genetic basis, we also now know that Huntington's is not only about the chorea and dementia. The chorea tends to be prominent early in the disease, but in the latter stages is replaced by a paucity and slowness of movement, balance issues and abnormal walking – something akin to Parkinson's disease. While dementia is seen later on, patients often experience problems with their thinking many years before other symptoms, with difficulties processing information, recognising emotions, and in attention and planning. Similarly, psychiatric manifestations are common – features such as anxiety, irritability, depression, obsessive-compulsive disorder and even psychosis – again often occurring many years before the more obvious onset of the disease. For Becky's father, it was the failure to recognise that Huntington's may affect behaviour long before chorea arises that led to his delayed diagnosis.

After their wedding, Rhett remembers he and Becky trying to help her father, getting him on medication, only for him to stop taking it after a while. 'It got to the point where he got really weird. The whole family had an intervention, got him down to Oklahoma City, where we were living. We finally got him diagnosed. He had forty-two repeats.'

By the time of his father-in-law's diagnosis, Rhett and Becky had had two children, a four-year-old and a newborn baby. Becky and her sisters got tested shortly after. Waiting for the results must have been terrifying; getting them must have been earth-shattering. Becky had forty-four repeats – two more repeats than her father – confirming that she had inherited the condition and demonstrating genetic anticipation in practice.

Rhett downplays getting the diagnosis. 'You know, we took it in our stride. My mother had just died of breast cancer,' he says. 'We knew not to get real excited about tests. A lot of people questioned it, but we had more kids knowing [that Becky was carrying the Huntington's gene]. I couldn't shake this feeling that there were people missing from our family. I think it was from God. We banked cord blood, thinking about stem cells.'

As Becky's father continued to deteriorate, they moved back to Kansas, but otherwise life carried on as before. 'If there were effective treatments, if there was some research to suggest there was something we could do, change our lifestyle for example, then maybe we would have done something else,'* says Rhett. 'I look at it a little bit different. None of us get out of here alive. Something is gonna get all of us. Heart disease, cancer. Huntington's is just one of those somethings. That is how I chose to look at it. The end game was the same. So, we just chose to live like that. The upside of getting the genetic test was that I knew what was going on, that we could plan our lives a little bit.'

I am a little taken aback at this view, although I am impressed by their courage. I think about the implications of receiving knowledge of a time bomb within your wife's genetic code, when she is in her mid-twenties; an awareness of the inevitability of a dreadful neurological disorder announcing itself in the coming years, and the risk that this condition might be passed on to your children. A toss of a coin, heads or tails. A 50 per cent risk for each child. But for Rhett and Becky, life simply marched on: 'By then we had four kids, and we might have had a fifth, but Becky was simply unable to carry [another pregnancy].'

That genetic time bomb did not simply explode, however. The fuse burnt long and slow. In hindsight there may have been signs of the disease from the early days of their marriage. 'We didn't live together before we got married,' Rhett explains. 'I knew after a couple of months that she didn't process information like other people did. But she was having some irritability [even] back then. It would come and go, just sporadic, and it was always associated with her period. And I thought: "Well, she is kind of like her dad," but I chalked it up to learned behaviour.'

By the time they moved back to Kansas this irritability had

---

* While treatments for the symptoms of Huntington's exist, there remain no current therapies available for the underlying disease. However, various trials for different types of gene-modifying or gene-blocking therapies have been undertaken or are ongoing.[18]

worsened, to the extent that Becky was started on an antidepressant.
'But it just made her into a zombie, and took away her sex
drive,' Rhett says. 'So we backed off the medicines. I reached
the point where I could deal with the mood swings.' He describes
periods of profound anger and aggression in her (reminiscent
of those individuals in Chapter 1, on wrath). Typically, Becky
would hurl accusations at him – of being a terrible husband,
of not caring about her, of not loving her. 'It's hard to take that
from somebody you love, even though by then we knew the test
[results]. Because those words, they're out there and they roll
around inside your head, even though you know that they're
coming from a place that's probably not her. It still affected
me,' he adds. Sometimes, the focus of her anger would have
repercussions for others, too. 'Occasionally, she would accuse
me of an affair, which obviously didn't occur. We had an
exchange student from Holland one time, and she thought I
was having an affair with her when she was a seventeen-year-old
girl. I was like: "I don't think so." She made her leave the house,
go live with another family. I mean, I felt terrible [for the student].'
At times, this would escalate to physical violence – Becky beating
him – although he says she never caused serious injury.

With the passage of time, the bad times became more
frequent, the mood changes more pronounced. Rhett found that
the only way to cope was to leave the house, to take more shifts
at the hospital some thirty miles away from home; night shifts,
weekends, a respite from Becky's disease: 'It was just easier to
be gone.'

Becky's irritability could to some extent be mitigated against.
But Rhett describes another symptom that was evident very
early on, one that is more relevant to sloth: a lack of motivation
to do anything. Complete and utter apathy. Indeed, apathy
seems to be the most common behavioural symptom in

Huntington's, and is present at every stage of the disease.[19] Apathy progresses consistently in tandem with the disorder's relentless march.[20] 'We would be packing the kids up to go some place, and she'd be sitting on the couch. My kids were seven or eight years old, trying to figure out how to pack their suitcase. She would just be sitting there telling them what to do. And then get mad at them when they wouldn't do it,' Rhett says. At other times, if their children were hungry, Becky would simply place a bag of crisps in front of them, despite previously trying to feed them a healthy diet. 'We eat organic, we [buy from] Whole Foods,' he tells me.

As with the mood fluctuations, over time, Becky's apathy intensified. Initially, Rhett recalls that if they were going out as a family, or going to church, Becky would still take pride in her appearance, putting on make-up and doing her hair. As the months passed, however, her lack of motivation worsened. 'I would come home from working a seventy-two-hour shift in the ER, and she would look terrible. She would be asleep on the couch, kids running around, the dogs having pooped on the floor. I would spend the whole day after I'd already been up seventy-two hours cleaning the house, getting meals prepared, going to the grocery store. She'd just be asleep on the couch, and then she would wake up and yell at me for being a bad husband.'

With the arrival of their fourth child, a daughter, the stark reality of what they as a family were facing crystallised. 'She would just forget about the baby. My nine-year-old son did most of the diaper changes and would get her out of the crib and bring her downstairs.' It was this lack of motivation that led Rhett to largely give up his clinical practice, uncertain of his wife's ability to keep herself and their children safe and fed. In recent years, Becky's capacity to look after herself or the children in any meaningful way has completely disappeared.

Interactions between the frontal lobes and other brain areas

If you were to study Becky's brain scans, you would likely see a very specific pattern of changes. The degeneration of the brain in Huntington's disease is not indiscriminate. If you look at the scans of people with the disorder, the appearances of the brain are often rather arresting. In normal brains, the fluid-filled chambers of the lateral ventricles in the frontal lobes are narrow – the basal ganglia constituting the wall of the cavity plump and bulging inward (Figure 1). In Huntington's disease, these deep structures of the brain are withered, causing the lateral ventricles to appear enlarged, their normal concave appearance rendered convex through the absence of brain tissue.

The structures that comprise the basal ganglia, which we encountered in the context of Parkinson's disease, have an important role in regulating motor function. It is not surprising that damage to these brain regions results in the movement abnormalities of Huntington's, as it does in Parkinson's disease too. But they have roles beyond just our movements. These deep brain structures are also intimately involved in communicating with the pre-frontal cortex, influencing decision-making and planning. They are also tightly linked to the limbic system, the major circuitry underlying our emotions, and are implicated in the self-generation of movement, action and volition, driven by emotional factors.

It might be expected then, that in conditions like Huntington's, Parkinson's disease or particular types of stroke which predominantly affect the basal ganglia, injury to these brain regions – not just the frontal lobes – influences the complex system that governs our decision to choose, to act, to expend effort in the search for reward. If the basal ganglia constitute a link between the process of decision-making and how our emotions influence those decisions, it is easy to understand how the basal ganglia may regulate our motivation. For example, specific damage within the basal ganglia from stroke has been linked with an extreme form of apathy, resulting in a complete loss of initiation of emotional or cognitive responses:

> Patients tend to remain quietly in the same place or position all day long, without speaking or taking any spontaneous initiative. When questioned, patients express the feeling that their mind is empty. The decreased number of spontaneous voluntary actions is clearly associated with a drastic drop in the number of the patient's daily activities.[21]

Curiously, these patients can be triggered into action through prompting. Direct questioning will precipitate appropriate responses, and if asked to do something, they will be able to. What is really striking is the disconnect between their ability to respond to the outside world and their complete inability to generate any actions from within. The machinery to act is present, but the inputs from drivers to act, like their emotions or measures of reward, are impaired.

In addition to their role in the regulation of movement, therefore, the basal ganglia appear to be critical for self-initiated actions. When it comes to this neural system that underlies motivation, it seems that the frontal lobes, particularly the pre-frontal cortex, are responsible for the generation of actions or behaviours, but are guided by internal drivers. Impairment of the frontal lobes is the cause for those with apathy in frontotemporal dementia, for example. The limbic system mediates the emotional component to these behaviours and their reward consequences. Yet it is the basal ganglia, the areas damaged in Huntington's, that mediate the activation of these brain regions, as the internal driver for our thoughts or actions[22] – as in Becky's case. Therefore, any alteration in any one of these circuits –

affecting our generation of choices, weighing up these choices, emotional responses to our actions or our internal drive to act – can give rise to apathy, which perhaps explains why it is so widely seen in neurological disorders of so many different types.

As I write this, it is inevitable that I indulge in some introspection. It makes me think of all the various factors that influence my own motivations. I have come to the realisation that I am intrinsically rather lazy. I dislike exercise, and generally will always seek the quickest option for any particular task, the easiest path. I like to frame it as 'efficiency', but in truth it is a form of sloth. When I look at myself, and the factors that push me to practise as a doctor, to write, to do all the other things in my life, I have to admit to myself that there is a maelstrom of emotions at play, not only the reward or pleasure of achievement: guilt, the expectations of myself and others, the need to feel productive and contribute. It illustrates the complexities of the various emotions that regulate my own behaviour, originating from my limbic system, mediated by my basal ganglia, and manifesting through my frontal lobes.

* * *

With the passage of time, the relentless progression of Huntington's disease has ravaged both Becky's brain and her family, devastating the woman Rhett found so mesmerising all those years ago. 'The thing about Becky, when we met,' he says, is that she 'made everything in my life beautiful. She was beautiful, the sweetest girl I'd ever met. She just had this quiet gentleness about her. All that's gone now. It got twisted into this. Something that's completely different.'

Becky's apathy has intensified to the extent that she does very little now. Aged forty-nine, her motivation to participate in the day-to-day acts of social engagement, of looking after her children, even of self-care, has been whittled away to nothing. 'She just sits there for days. You've seen patients in a nursing home,' Rhett tells me. 'And you know how they are in a nursing home; they're sitting there in a chair watching TV, right? And then you come by, and you say it's time to eat. And you bring over the table, and you feed them. That's what I'm doing.' A housekeeper comes several times a week to help Rhett look after his wife and children, while he tends to the ranch and still does occasional shifts in the local hospital. He adds: 'At this point, there is not a lot of enjoyment' – although I am uncertain if he is talking about Becky's life or his.

I broach the issue of their current relationship tentatively. Rhett starts by saying he is a very different man now to the one he was in his twenties. His faith has been an enormous source of comfort to him but has also cemented his role in his own mind. 'There's a verse in the Bible that talks about marital relations, and it says, that the man has to love his wife as Christ loved the Church and gave himself up for the Church. And so I had this picture of Christ on the cross for me. And so I could do no less for her.' It is also Rhett's own upbringing that has guided his actions. Rhett was brought up in a divorced household, his mother marrying three times. 'I grew up with step-parents and I didn't want that for my children. I looked at my role. I was to shield the kids as much as possible, to provide stability for them, but also take care of her [Becky]. I think our culture, our society, especially in America, we're idiots over here sometimes you know, we're so short-sighted. The only one person to gain from a divorce or to leave her would have been me.' His own motivations are a complex mix of feelings: duty, faith and protecting his wider family.

Despite Rhett's desire to shield the children, they have not remained unscathed. With each child having a 50/50 chance of developing the disease, the spectre of Huntington's hangs over them. Rhett is surprisingly phlegmatic. 'I've told my kids: "Look, my mom was dead at fifty-six of breast cancer. My father-in-law died at sixty-five of Huntington's, your mother's aunt died at fifty-six or fifty-seven of Huntington's. People have heart attacks and die at forty, or they have a car accident." What I've tried to teach my kids is: live every day to the fullest. Something will get you, it's just a matter of time. And so maybe it's a bit fatalistic, but I've encouraged them to live their life.' I ask him if any of his children have had a genetic test to ascertain if they carry the Huntington's disease gene. Initially, he thought that the family should seek as much information as possible, but his views have changed. 'There's not an effective treatment at this time. Why put that burden on yourself at this point? I've always maintained with the kids to go live life, because none of us are guaranteed tomorrow. My son is going to get married; we've talked with my [future] daughter-in-law, and she's well aware of it. My son-in-law – I talked with him before they got married – "Are you aware of this? You see what's going on, you see how your mother-in-law is acting." He's going to be a pastor. So he's kind of taken my stance – "I'm gonna love her." I've told the kids that it is up to them if they want to get tested.'

Rhett describes a double-edged sword when it comes to his wife's own genetic test. 'It was a blessing and a curse, because I symptom-hunted. I was the worst symptom hunter in the world, but I was medical, I was trained to look for symptoms. It finally occurred to me not everything's Huntington's, but everything is Huntington's. But having the knowledge was also good because I was able to plan a little bit. Had I not had the knowledge, had we not known about her dad or her, honestly I probably would have left her if she wouldn't have been sick, because she got so difficult.' For a man like Rhett, such an admission must be incredibly painful. 'Even with my faith, I'd probably ask God to forgive me, because I just can't do it anymore. But when I realised she was sick – I mean, I did say in sickness and in health, you know. So, yeah, do I want to [look after her] every day? No, absolutely not. I mean, there's still a part of me that wants to be twenty-two. And go hang out with a Swedish bikini team, or whatever.'

This phrase, 'Not everything is Huntington's, but everything is Huntington's', is one that Rhett repeats several times in our conversation. He uses this to convey his view that while anger, irritability, apathy and all the other symptoms of Huntington's are experienced in normal life, in Becky and others who have this condition, the world is mediated through a diseased brain. For Becky, Huntington's is the filter through which all of life is conveyed.

Understanding the precise nature of that filter, exactly what Becky's current world looks like, is a challenge though. Even in day-to-day life, we suppose – rightly or wrongly – that the minds of others are similar to our own; have the same basic machinations, thought processes and emotions. As is so often the case in neurology, however, in the context of a diseased brain, such assumptions are even more problematic. We are even more reliant on our own observation, on extrapolation, and more crucially on those who live alongside the neurologically altered, who know them and love them. To comprehend how Becky's Huntington's brain shapes her perception of her life and world, we are almost entirely dependent on Rhett's thoughts and words. And these too percolate through his own mind, moulded by his own brain, his own character.

* * *

This carefully choreographed dance between the various circuits of the brain ultimately asks one specific question of our bodies: is it worth it? Is the effort of expending energy in the pursuit of a reward, its size and the likelihood of obtaining it, merited? The generation of different options, the evaluation of the payback, the calculation of the possibility of achieving the goal – all these are important elements in the motivational process. But there is another factor is this equation. We also need to measure effort, to have a metric of exertion, to calculate the physical or mental cost of any particular action. And it is this particular aspect that may give rise to a facet of 'slothfulness' beyond those individuals like Becky with Huntington's, or those with Parkinson's or frontotemporal dementia. A misperception of effort, an overappraisal of exertion, may in part explain why many people, with medical conditions or otherwise, ex-perience fatigue.

It would be difficult to describe AJ as slothful. She is currently undertaking an undergraduate degree at the age of forty. In her prior life, she has run several businesses. 'I have actually been an entrepreneur since I was sixteen,' she tells me. 'My father believed that because he got working so early, it meant that I ought to. So, I tried going to work with a few of my friends, but without any real conviction. All of it looked pretty terrible to me. I had a friend who was a waitress, who got constantly sexually harassed. There were friends of mine working in the fields – absolutely gruelling. So instead of getting a job, I made a job.'[23]

While still in high school, AJ started a shop focused on spiritual pursuits, multiple religions, multiple cultures, indeed counterculture. Growing up in a conservative small town in Bible Belt Ontario, she says she always felt a bit of an outsider. On her mother's side, she is the descendant of Jenische immi-grants – an itinerant people originating from Central and Eastern Europe, often conflated with the Roma people, but quite distinct – and felt intrinsically drawn to non-mainstream cultures. 'Jenische has an inherent mix of Western and Eastern

European cultures, some aspects of paganism, Catholicism and Judaism. My father was born in Toronto and doesn't have a whole lot of culture. So, it was very easy to identify more with a richer culture [from my mother's side], to be honest.'

The shop did very well from early on, but AJ describes it as 'outing' her as other, creating friction at school. 'I was encountering difficulties with teachers and the community. So I just said: "Screw this, I'm making money, goodbye." I ended up quitting school.' She started her working life, of running shops, online businesses and other employment.

While AJ had a strong work ethic from a young age, she could not be accused of being physically slothful either. A severe persistent headache after a bang to the head in her late twenties led her to gain a lot of weight. After suffering headaches for years, it was eventually discovered that she had developed a rare condition: idiopathic intracranial hypertension. The headache attributed to her concussion had been replaced by pain due to increased pressure in the fluid around her brain – a disorder of uncertain origin but often associated with weight gain in young women. Rather than taking medication, AJ took the harder option. Intense regular exercise, coupled with dieting, allowed her to lose over 45 kg (100 lbs), and the idiopathic intracranial hypertension resolved. 'I ended up quite a fitness nut. I could do two hours on an elliptical [training machine],' she adds.

About a year after her recovery, her marriage of five years fell apart. She moved back in with her parents, at her mother's behest. Her father had been diagnosed with brain cancer, and her mother was struggling to cope after a year of caring for him at home.

It was in an effort to ease her mother's load that AJ's life changed beyond recognition. Noticing that her mother's increasing frailty was an obstacle to maintaining the garden, AJ took action. 'I decided to fix up the garden,' she says.

Her handiwork did not quite go to plan. Instead of a successful remodelling of the garden, she was rewarded with a catalogue of injuries. 'I ended up hitting my head on – of all effing things – a birdhouse. That same day, though, I went to get more wood for the project. I stupidly said to the people that would have loaded it: "Oh, no, it's okay. I've got it.' And ended up hitting my [head] on the hatchback [of the car]. And then after that I unloaded the wood, and I ended up saying: "You know what? I feel like crap. Today has sucked. I'm gonna go to the pool." While I was swimming in the pool, well, I guess I was somewhat already disoriented. It ended up I was swimming on my back and I ended up hitting the edge of the pool with my head.' But this was not the end of AJ's bad day. 'It gets worse. And then here's the big one. I get out of the pool, I go into the changing room, I slip and I fall onto the concrete. And that was definitely a full-on new concussion. Stupidly I did not go and seek treatment. I just [thought]: "Oh my God, I'm going to bed." And I ended up sleeping almost continuously for days and then it was a couple of weeks later that I realised: "Oh shit, this isn't good."'

In the aftermath of a succession of blows to the head, AJ was floored. When she managed to get out of bed, she was troubled by other symptoms too. 'I very quickly noticed things like frustration and anger. And also difficulty with processing sensory inputs,' she says. 'Everything seemed way too loud, in a way that was inherently irritating. My focus was really trashed. I couldn't cut through ambient noise to hear a conversation properly. If I sat down to read, even a long sentence, by the time I was at the end [of the sentence], I would forget the beginning.'

Many of AJ's symptoms settled over the subsequent weeks, but two aspects of her health did not. One was her sleep. 'Honestly, I was still sleeping way more than I would call normal for at least a year. [In the few weeks after the concussion] if I was awake even eight hours, that would be surprising.' Even while awake, the urge to sleep was irresistible. 'I would fight to keep my eyes open, to remain awake during that time.'

The second aspect was fatigue, although initially distinguishing between the sleepiness and fatigue was difficult. As her sleep gradually improved over the months, AJ became more aware of this as a separate issue. Despite no longer feeling quite so sleepy, her energy levels were completely depleted. 'I felt heavy and tired at all times. There was no amount of sleep that was sufficient.' That fatigue was clearly physical: 'Even doing a normal chore, like the laundry. Right after that exertion, I was like: "Oh my God, I need to rest. I can't do the next thing."' Her ability to exercise evaporated. From spending two hours on a training machine, she could barely do anything. And any physical exertion would have repercussions. 'Even doing the groceries. I would not only be exhausted for the rest of the day, but the next day I would still feel the exhaustion.' I ask her if she just felt weak, or if her muscles ached. 'There is a certain kind of muscle ache to it. But it is not the same as when you work out. It felt almost anaerobic, like you're deprived of oxygen. If I try to lift a case of water, for example, I remember what it was like before this, okay, it was heavy, but now I feel the exhaustion, the weight of it.'

This fatigue extended into the mental realm too. Previously a big reader, AJ found it exhausting to read a book, even a sentence. 'It was hellish. Really demoralising. Even after a year, as things improved a little. I used to be a major non-fiction reader, but I couldn't connect the ideas and terms. It was exhausting to read that kind of work.' Her medical team advised her to read *Harry Potter* instead.

* * *

As AJ demonstrates, tiredness can come in many shapes, and those forms are often conflated or overlap. Various terms such as sleepiness, tiredness and fatigue are frequently used interchangeably, but in a medical setting have specific and different meanings. Excessive sleepiness describes the inability to stay awake, to resist sleep, while some patients have an excessive need for sleep, usually defined by a requirement of over ten hours' sleep within a twenty-four-hour period. There is even a term for prolonged periods of time spent in bed, despite no objective evidence of an increased need for sleep: clinophilia. In contrast, fatigue has a very different meaning, being defined by physical or mental exhaustion associated with difficulties initiating or sustaining voluntary activities. Critically, fatigue is not substantially improved by increased rest or sleep.[24]

Unfortunately for her, AJ has faced almost all the above: the very prolonged sleep requirement, the sleepiness when awake, and the mental and physical fatigue. A toxic mix of tiredness that has influenced almost all aspects of her life; a 'sloth' that has no moral significance.

She is not alone in her experiences after a head injury, however. In fact, both sleep disturbance and fatigue are incredibly common after even very mild head injury. A prolonged need for sleep and daytime sleepiness are well recognised as consequences, but fatigue is the most frequent complaint, even ten years after traumatic brain injury (TBI), affecting up to 80 per cent of survivors.[25] The cause of sleepiness is thought to relate to direct damage to those regions in the brain critical to regulating sleep and maintaining wakefulness, but the origins of post-traumatic fatigue are less obvious and more controversial.

Is physical fatigue just exhausted muscles?

It seems logical that our perception of physical effort might be related to signals from exhausted muscles. That physical fatigue or perception of exertion with intense physical activity relates to sensory feedback from those muscles. Sensory information from an organ involved in effort, usually a muscle, must feed back to the brain to generate such perception of exertion.

However, if physical exertion is measured by the activity of our muscles, and fatigue is purely a function of how much our muscles have worked, explaining certain observations is very difficult indeed. It does not rationalise how, under particular circumstances, we can display superhuman levels of strength. There are many examples of individuals who, under stressful or highly emotional circumstances, have exhibited strength or exertion that is almost inexplicable. People like Zac Clark, a sixteen-year-old Ohio teenager, who in 2019 was gardening at home with his mother when they heard a neighbour crying out for help. When they ran to investigate, they found the neighbour's husband pinned under their family car, a VW Passat. His legs were free, but his upper body was being crushed by the weight of the car, the jack having given way as he worked under it. In the heat of the moment, without thinking, Zac singlehandedly lifted the car off his neighbour, allowing him to be pulled out, and undoubtedly saving his life. Or Lydia Angyiou, a forty-one-year-old mother living in a remote Quebec village. As she was walking with her sons, a 320 kg (700 lb) polar bear attacked them, and she stood between it and her sons, wrestling the bear until a neighbour fired shots in the air and it ran away. Her maternal instinct left her with only cuts and bruises.

These kinds of cases, termed hysterical strength, illustrate that under stressful life-threatening circumstances, our ability to exert ourselves, to suppress physical fatigue, cannot be underestimated.

They show that our emotional state can dramatically shift those processes that limit the work our muscles can usually perform, and hint that these mechanisms must involve the brain as well.

In AJ's case too, her description of physical fatigue is one of profound exertion to perform tasks that should be very untaxing. It is problematic to link her physical fatigue to a change in muscle activity or sensory signals. It is a head injury that seems to have triggered her symptoms. Her situation marries with this view that the origins of physical exertion are indeed the brain.

And beyond these cases of hysterical strength and people like AJ, additional evidence has also cast significant doubt on a view of physical fatigue purely emanating from the muscles.[26] For example, some substances like botulinum toxin (Botox) or curare weaken muscles without affecting sensory feedback. These kinds of drugs therefore limit the work that the muscles can do, and would be expected to lessen our perception of exertion if fatigue were simply a function of muscular contraction. Instead, people given these agents experience much more fatigue, rather than less.

Epidural anaesthesia, where a local anaesthetic is injected around the nerve roots as they exit the spinal cord, blocks sensory signals from the skin, muscles and other tissues from returning to the brain. This sort of anaesthesia might be expected to reduce the sensation of fatigue, since exertion signals from the muscles would be impaired. Instead, if you get people with an epidural to cycle on a machine, their sensation of physical exertion is either unchanged or actually increased.

These findings, among others, have led researchers to conclude that physical fatigue does not emanate from the periphery – regions outside the central nervous system. That instead, the perception of fatigue or physical exertion is almost entirely within the brain itself.

If this is the case, that the work our muscles are actually

doing has little or nothing to do with our perception of effort, how might this happen? How else does the brain measure exertion, in the absence of this information from our muscles? This would require the brain to inform itself how much it is asking of the body, without the need for any additional knowledge. And indeed, it seems that this is precisely what it does.

Movement of our arms or legs results from the firing of neurones in the primary motor areas of the cerebral cortex – the arm and leg regions of the motor strip located in the area of frontal cortex closest to the back of our skulls (Figure 3). These impulses are transmitted down our spinal cords, through our nerves and directly to the muscles themselves, making them contract. These impulses do not only reach our muscles, however. Simultaneously, these motor areas send signals elsewhere, to other regions of the brain. These additional messages, termed corollary discharges, inform regions of the cerebral cortex responsible for sensation that muscles are being activated. These sensory areas do not need any information from the muscles to know that they are working hard. They are directly informed by the control centre. This would certainly explain why muscles weakened through toxins result in increases in fatigue rather than decreases: the increased stimulation of these muscles from the motor areas of the cerebral cortex to achieve the same motor output would be perceived as increased effort.

Assuming the latter explanation, that of corollary discharges, is correct, then the question remains as to how exactly these motor signals to sensory regions of the brain are conducted. Functional MRI studies have implicated an area of the motor cortex in this process: the supplementary motor area (SMA). Activity in the SMA is directly proportional to physical activity intensity, and the SMA has direct connections to the sensory cortex.

Astonishingly, it is possible to lessen physical fatigue by zapping the SMA with a powerful magnet. In one remarkable study, twelve healthy volunteers were asked to squeeze an instrument called a dynamometer (an instrument that measures grip strength) in their right hand. During this task, a strong magnetic field was applied to a localised area of their brains, either to the primary motor area, the SMA or a control region of the brain – an area thought to have no role in movement or sensation. This magnetic pulse causes a very small area of the cerebral cortex to transiently work abnormally, briefly disrupting normal neurological function. Disrupting SMA function, but not primary motor area or the control region, during the task dramatically reduced the subjects' perception of exertion, presumably through blocking the normal output signals of the SMA to the sensory cortex.*[28]

The relationship between physical and mental fatigue – a unified theory

The brain therefore has its own internal monitoring system, to assess exertion and its costs, and disrupting this system overcomes physical fatigue. So what about cognitive fatigue, the decline in

---

*   The SMA is not the only brain region associated with physical fatigue or meas-ures of exertion. Other brain areas have also been implicated – regions like the anterior insula, for example. Like the SMA, this location in the brain is linked to motivation, and the assessment of effort or exertion. The insula is a region of cortex folded deep within the fissures of the brain, sitting at the junction between the frontal, temporal and parietal lobes. Its anatomical location provides a clue as to the complex nature of its functions, linking emotions, self-awareness, the processing and integration of our multiple senses, even consciousness.

   Among its myriad functions, the insula is crucial to monitoring of our internal state, signals arising from within our own bodies, termed interoception. For example, the structure and function of the anterior insula has been shown to correlate to people's accuracy in determining their own heartbeats.[27] It is not just important for monitoring these physiological signals, however. The insula is thought to have a broader role in our awareness of state, when it comes to thirst, hunger or pain. Obviously, an ability to detect what our own bodies are doing or feeling must be fundamental to our perception of physical exertion and associated physical fatigue, but also interpreting exertion in the context of the needs of our own bodies.

mental performance with cognitive rather than physical tasks? AJ of course has both types of fatigue, finding both physical and mental exertion deeply exhausting.

There is evidence for the role of this motivational and monitoring network in cognitive fatigue too. In one experiment, nineteen healthy participants were asked to perform a pretty boring task, involving monitoring a screen for cues that instructed whether to click a button with their left or right hand.[29] Over the first two hours of the task, predictably, the speed and accuracy of the participants declined. At this point, they were offered some additional motivation in the form of a financial reward if they performed well for the remaining twenty minutes of the task. This potential reward significantly improved task performance, either through improved accuracy or reaction time. The implication is that cognitive fatigue can to some extent be overcome by the hope of reward.

Another study also demonstrated the influence of cognitive fatigue on our perception of reward. Individuals were asked to perform a task requiring attention and memory for about six hours in total. As the task proceeded, the participants began to favour small, immediate financial rewards over larger delayed ones. This change in reward evaluation was mirrored by a decrease in activity in the lateral pre-frontal cortex on brain imaging.[30]

Taken together, these two studies therefore suggest a strong link between motivation and mental fatigue too. That fatigue can be overcome by the perception of increased reward, and that mental exertion alters the way in which we perceive and evaluate reward.

Indeed a battery of neurobiological studies have demonstrated that those areas of the brain implicated in physical fatigue are also relevant to cognitive fatigue.[31] This does not only point towards evaluation of reward having a role in cognitive fatigue.

As with physical exertion, these studies also hint that the monitoring of cognitive exertion may have a bearing in cognitive fatigue as well. For example, the SMA shows changes in activity in the mental fatigue seen in anhedonia, the loss of pleasure associated with depression.

These various brain regions together monitor a range of inputs, our internal body state, signals from our muscles and organs, and systems involved in our cognition. Utilising all of this information, and the value of any reward, these regions then influence the degree of motivation to try, to attempt to obtain a goal, and crucially: our perception of fatigue.

Again, the question that we are asking of our bodies is: is it worth it? The phenomenon of fatigue may originate from changes in the function of any of numerous circuits within the brain – those regions that monitor our movements, those that scrutinise our internal state and those that determine motivation. Any imbalance in these widespread and disparate yet connected systems can give rise to fatigue, both mental and physical. This is a potential explanation for why fatigue is so common in people with neurological and psychiatric disorders, and indeed in the general population.

This view of fatigue – that it emanates from the brain – is not mutually exclusive to other interpretations of fatigue. For example, chronic inflammation may be responsible. As we have already heard, infection can induce lethargy, a lack of motivation and an array of other physiological and psychological functions. Fatigue is also incredibly common in a range of autoimmune disorders, such as rheumatoid arthritis and lupus, where the immune system causes widespread inflammation throughout the body.[32] Specific brain cells mediating these changes have been identified, in mice at least, as we have seen. From an evolutionary perspective, it makes perfect sense that having inflammation of any sort within the body, due to damage or infection, should generate fatigue. Encouraging the body to rest, to minimise exertion while ill, to preserve energy and reallocate it to the immune system, would facilitate healing.

In some conditions, certain markers of inflammation, called cytokines, are elevated, leading to heightened fatigue and a reduction in effort. These cytokines have been shown to alter activity within the basal ganglia,[33] and inflammation anywhere within our bodies may profoundly influence those chemical systems that mediate the evaluation of effort versus reward. This may also go some way to explain fatigue and apathy in psychiatric disorders such as depression, which in recent years has been increasingly linked to inflammation within the body.[34]

Ultimately then, fatigue is largely a function of the brain. However, it is not purely a consequence of exertion. It is also directly influenced by internal body signals, and importantly by the brain circuitry that evaluates reward versus effort, that defines motivation. It makes me think of athletes who listen to heavy metal music before a race, using anger to boost motivation and overcome fatigue. Or those cases of apparent superhuman strength in the face of mortal danger, like Zac Clark lifting a car, or Lydia Angyiou protecting her sons from a polar bear, where the reward outweighs any measure of effort.

While apathy, anhedonia, physical and mental fatigue are not entirely the same entities, there is a huge degree of overlap, in terms of their outward manifestations, and their behavioural and neurobiological bases. They correlate highly, both in healthy individuals and those with neurological or psychiatric illness. At their core are the brain circuits that influence motivation and effort. In certain circumstances, such as physical injury, inflammation or infection, a fleeting attenuation of our drive to do, to achieve, to win, can be helpful and can aid our survival. If these brain networks are impaired by less transient issues, then this gives rise to 'sloth', and all its implications.

The merging of the 'mental' and 'physical'

As with other areas of neuroscience then, in the world of fatigue the 'mental' and the 'physical' become less distinct. There are ways to explain fatigue in both psychological and physical terms, with an underlying basis in neurobiology. What about in AJ's case, where her repeated head injuries have precipitated a debilitating 'sloth'?

In TBI, some researchers have argued that the injured brain needs to work harder to compensate for any impairments in attention or processing speed, making performing the same task intrinsically more mentally arduous.[35] The fatigue that people with TBI experience has been found in some studies to be independent of factors such as depression, pain or poor sleep, and authors of these studies conclude that the fatigue in part is attributable to brain injury itself.[36]

Other researchers, however, point out that fatigue is often entirely unrelated to the severity of the injury. They argue that this 'post-concussion syndrome' may result from complex interactions between functional and structural changes to the brain, a genetic predisposition and psychosocial factors. They point to other studies that show the association of fatigue with depression, anxiety and post-traumatic stress disorder (PTSD), as well as psychological factors such as negative beliefs about recovery, and increased vigilance for symptoms.[37] But as we have already seen in previous chapters, these 'psychological factors' may also directly influence brain structure and function.

AJ herself recognises the possibility that psychological factors may play a role. She describes her first relationship as a teenager as being violent, manipulative and abusive, of being the victim of repeated sexual assaults. She has previously been diagnosed with PTSD. She tells me: 'PTSD changes the brain; it affects the way you deal with stress. It does in and of itself cause some fatigue.' But she is reluctant to ascribe all her fatigue to her PTSD. She points out that in earlier years her headache caused by idiopathic intracranial hypertension was also dismissed by doctors as 'psychosomatic', attributable to her PTSD.

For me, this kind of division between physical and psychological factors is often deeply unhelpful in clinical practice. On a scientific level, there is the issue as to how psychological factors influence our physiology. As we have seen, this separation is artificial, since our psychology and physiology have shared origins. Our psychology emanates from the brain and can influence the brain. Our brains and minds are the same thing.

On a practical level too, there are problems with this categorisation of disease and dysfunction. Patients often recoil at the suggestion that their problems might originate in the mind rather than in the body. There seems to be a pecking order of causation, where 'physical' processes are considered real, and 'psychological' processes are somehow imagined or made up. These views taint clinical medicine and wider society. At the mere suggestion of psychological therapy, some patients bridle, affronted by the inference that I am telling them that their symptoms are 'all in the head'.

From my own perspective, it is a terrible pity that these views are so ingrained, and that the edifice of medicine has fuelled them. I recall twenty-five years ago seeing a very senior doctor pushing a patient paralysed from the waist down out of the emergency department in a wheelchair, abandoning him on the pavement outside the hospital, telling him not to waste his time.

There was no structural cause for his symptoms, no stroke or tumour in his spinal cord, but this did not make the poor man's weakness any less real. What was then called 'psychosomatic' would now be called functional neurological disorder.

In many of the complex cases I see, my interest is not in ascribing a 'physical' or 'mental' cause to a particular symptom or disease. My focus is on alighting upon a treatment or treatments that work, and making the individual sitting in front of me better. A drug can treat a 'physical' or 'mental' condition, but so too can psychological therapy. This division makes little sense scientifically, but it also hampers clinical medicine, and limits therapeutic options.

Even now, several years later, AJ's energy levels remain affected. 'It feels as though it comes from an extremely taxed physiology. You are constantly defending what little focus and energy you have.' She describes a need to carefully balance energy expenditure, to budget for the days ahead. As if her batteries do not charge properly, or empty faster than everyone else's. 'I basically budget for one social activity a week. Even to go out for a few hours is extremely depleting. I have come to understand the cost to my body. I have learnt to [consider energy] as a resource. That is a learned skill.' She tries to preserve herself for her degree. 'I realise that my energy levels are nowhere near normal. My functioning day-to-day is about 30 or 40 per cent of what it was.'

In my own mind, it is not AJ's physiology that is taxed. The flesh and bone of her body, the muscles that do the work, are unaffected. It is her brain's assessment of that work that has been profoundly altered, amplified by the damage that has resulted from her head injuries.

One of the challenges for AJ has been to communicate her symptoms to others. 'It is really difficult to describe to people.

Honestly, since Covid, it has become easier, because I have found that more people have experienced it now. Like when you are really sick. Your body is functioning. You can get up and make tea. But the idea of actually concentrating, doing work, doing anything. It is not even that you are just exhausted. It is: "How can I get my brain to do this?" There is something in the way, something obscuring.'

Yet with time that veil of fatigue is very gradually lifting, her energy levels inching up month by month. And despite her symptoms, she is now pursuing the education she abandoned for working life: a university degree, while carefully walking the tightrope of energy reserve and expenditure.

* * *

I ask Rhett what his and his family's views are on pre-implantation screening, the process of assessing if an embryo carries a disease, and only implanting those embryos that are free of the disease-causing gene. His response is unexpected, more nuanced than I would have predicted. 'It's a really tough thing to answer. Would I criticise anybody for doing it? Absolutely not. Not in a million years. Everybody's got to come to that decision them-selves. I personally believe there's some verses in the Bible that indicate life begins at conception. And so God loves all people, just how they are. I mean, if he creates somebody with Down syndrome, he does it for a purpose. Who can know the mind of God? But, if my kids chose to do that, I'm probably going to keep my mouth shut.'

While on the subject of religion, I am fascinated by how Rhett's Christian faith has been influenced by his first-hand witnessing of neurological disease producing 'sinful' behaviours – Huntington's generating sloth/apathy and wrath in Becky – and whether this has had broader implications for how he perceives human behaviour. I point out that when he talks about Becky, he talks about her essence, and then the Huntington's, and ascribes some of her behaviour directly to the disease rather than it being a reflection of her 'soul'. Is that perhaps the case for others less obviously affected by neurological disease, but whose 'sinful' behaviours are simply a reflection of how their brains work?

Rhett's answer draws heavily on his faith. 'According to Christian teaching, we're all sinful. I'm no better than anybody else on this planet, period. True Christianity brings humility, a brokenness before God asking for forgiveness,' he says. 'You know, part of my behaviour in my twenties was due to the fact that my folks divorced. But that doesn't excuse the fact that I still did wrong, that I objectified women in my twenties, when they were just there for my pleasure, it doesn't excuse that one still needs to seek forgiveness from God and forgiveness from those that we've hurt.' For most people, Rhett implies, sinful behaviour, while it may have explanations, still carries with it responsibility, guilt and the need for forgiveness. Yet, he continues: 'You know, Huntington's is a bit different, because it destroys the brain. [In contrast] most people process pretty well. But when I can see something like my wife, who's an amazing, beautiful woman, get twisted into something that she's not through a disease process, I have compassion for those who are struggling.'

This is where Rhett and I diverge most obviously. Not only in his piety and my atheism, but also in what constitutes 'normal'. Rhett articulates well the sensible and intuitive view that a diseased brain is not the same as a normal healthy brain. Huntington's disease is very obviously not within this spectrum of normal. The world understood by a brain gnawed away by dementia is very obviously different to that of a healthy brain. But who is arbiter of what normal is? Is it a brain that looks normal on a brain scan? We know of many neurological disorders that produce normal-looking scans. Is a person whose behaviour is affected by severe physical or psychological trauma normal? Is someone who inherits a set of genes that influences their behaviour negatively normal? The principal issue is the dividing line: where normal ends and abnormality begins, where morality is suddenly replaced by pathology.

This boundary, between what constitutes the ordinary of the human experience, and what lies outside it, also extends to sloth. For sloth is usually a healthy trait, a balancing of the weighing scales of energy and reward, a preservation of resources for those actions crucial to our existence. For all of us, this balancing act is influenced by how we perceive reward, how we evaluate energy, and those neurological systems that underlie these calculations. It is just that for some of us, the point at which the balance is achieved is different. The presence of disease, either neurological, psychiatric or of the body, acts like a finger on the scales, tipping them in one direction. Normal sloth, heightened to extremes.

When we talk of pleasure and reward, we tend to consider them a goal. We speak of rich lives, of living for the moment, enjoying the brief time we have on this earth. But pleasure is simply a tool of evolution, the emotion that motivates us in the act of survival – to eat, to drink, to procreate, to seek out experiences that ultimately promote a longer life and the passing on of our genes. I think of those patients I have seen with intense apathy; or those individuals in the deepest pits of depression and anhedonia, left without the incentive to speak, or even eat or drink, needing electroconvulsive therapy as a life-saving procedure. What is life-threatening is not the loss of pleasure or reward, but the far-reaching consequences of that loss. Pleasure is the means to an end.

These brain processes that tally purpose and effort are the bedrock of every aspect of our being.

6.

Greed

In an unknown room somewhere in the world, an anonymous person drafted an email and clicked send. The entire message was only eight words long. When Bastian Obermayer, a journalist at *Süddeutsche Zeitung*, one of the largest German daily newspapers, opened his inbox, the email simply said: 'Hello. This is John Doe. Interested in data?'[1] The brevity of that message held no clue as to the volume or significance of the information to follow. Obermayer's curiosity was piqued, and his response to that message culminated in one of the largest ever leaks in journalistic history. So huge was the digital trove subsequently delivered to the newspaper – 11.5 million documents and 2.6 terabytes of data – that Obermayer and his colleagues enlisted the help of the International Consortium of Investigative Journalists to make sense of it all.

Over the next year, more than 600 journalists from 107

different media organisations spread over 80 countries pored over this vast collection and pieced together the financial arrangements of more than 210,000 offshore companies in 21 jurisdictions. Their initial report was published on 3 April 2016. The leaked documents constituted the records of Mossack Fonseca, a Panama-based law firm and provider of corporate services, and detailed the efforts of thousands of individuals, among the most powerful and wealthy people on the planet, to evade tax. It revealed an extensive and complex web of offshore shell companies, used as financial vehicles by the already extremely rich to avoid tax and get richer. Heads of state, politicians, celebrities and business people were all implicated. The great majority had not engaged in criminality. US President Barack Obama stated: 'It's not that they're breaking the laws. It's that the laws are so poorly designed that they allow people, if they've got enough lawyers and enough accountants, to wiggle out of responsibilities that ordinary citizens are having to abide by.'[2] Despite the lack of criminal wrong-doing, governments and electorates were scandalised. The leaked information resulted in many countries litigating against their tax residents, claiming billions in back-taxes and fines, and brought down politicians, including Iceland's prime minister, who had sheltered money offshore without declaring it.

A few of those named in the 'Panama Papers' did meet the threshold for criminality, however. Nawaz Sharif, the former Prime Minister of Pakistan, was fined over $10 million and sentenced to ten years in prison, although he fled to London, claiming his conviction had been politically motivated.[3] His conviction was subsequently overturned. Swiss bankers were convicted and fined hundreds of thousands of francs for moving millions of dollars of Vladimir Putin's friends and allies through this network of companies. A US accountant was sentenced to thirty-nine months in jail for wire fraud, tax fraud, money-laundering, and other charges. But for almost all those caught up in the Panama Papers, there was little punishment, except for the disapproval of their peers, electorates or subjects. Arguably, one of the people who suffered most was 'John Doe', who in a rare interview with the German news outlet *Der Spiegel*, said that they feared for their life, apprehensive of retribution for their role in the exposure of the rich, powerful and dangerous.

The leak of the Panama Papers, and the world's response to them, neatly encapsulates the ambivalence of our society towards greed. Of course, there was moral outrage that the richest among us would seek to avoid paying their 'fair share', going to such efforts to evade or avoid tax. That elected officials could display such hypocrisy, paying lip service to concepts of fairness and social justice, while simultaneously weaselling their way through international banking practices to limit their liability. World leaders preached from their lecterns that the global financial regulations needed overhaul, yet such change has proved painfully slow and limited. Undoubtedly there was also a degree of resignation, even acceptance, that this is what the rich do. I suspect in some corners, people might have even considered that such efforts to avoid tax are praiseworthy, provided the letter of the law is not broken. That this was an issue of poor legislation, not of moral failings – a normalisation of greed. But I also suspect that not a single person thought that those people implicated were physically or mentally ill, that greed can be a function of disease.

* * *

Search the medical databases for 'greed' or 'avarice' and 'neurology' or 'psychiatry', and there is almost nothing to find. Greed eludes our medical lexicon in a way that the other 'sins' do not.

For these other sins, we delineate the pathologies that shape our thoughts and behaviours, and set them apart from those underlying character traits through their intensity and consequences. For greed, we do no such thing. Yet greed, like the other sins, is perilous in its most extreme forms, causing harm to individuals and wider society alike.

Perhaps it is that greed is a mirror reflecting the values of society, or indeed is the foundation of our society. That while we accept it is bad, even as greed burns at its most intense, it is the engine of human ingenuity and progress, a potential force for the betterment of humanity. 'Capitalism has been called a system of greed – yet it is the system that raised the standard of living of its poorest citizens to heights no collectivist system has ever begun to equal, and no tribal gang can conceive of,' wrote Ayn Rand. Greed is the cornerstone of economic growth and innovation, of consumerism and economic activity. Or, in the words of Gordon Gekko in the 1987 film *Wall Street*: 'The point is, ladies and gentlemen, that greed, for want of a better word, is good. Greed works. Greed clarifies, cuts through and captures the essence of the evolutionary spirit. Greed . . . has marked the upward surge of mankind.'[4] Many economists consider greed as the driver of economic growth and development. This view of greed is not a recent one either. The Athenian historian Thucydides (460–395 BCE) wrote that greed is not unarguably negative, agreeing with Gekko that it motivates progress.

And perhaps it is because of this that we take a more nuanced view of this sin, the excessive or insatiable desire for more that is greed or avarice. More money, more stuff, more power, more influence. More of anything that is valued. Because what one person may perceive as avarice, another may deem ambition; what some may view as excess, others may label prudence. To pathologise it would jeopardise not only the essence of the society we live in, but also risk the imposition of one's own moral prism upon others, something most doctors are keen to avoid.

While the medical profession might exhibit reluctance in doing so, others are less hesitant. For each Ayn Rand or Gordon Gekko, there is the opposing view. Throughout human history, greed has been disapproved of, a trait harmful to society, resulting in unfairness, accrual of resources at the expense of others, selfishness and corruption. According to Saint Paul, *radix omnium malorum avaritia* ('the root of all evil is avarice'), and as Jesus is quoted in the Gospels: 'It is easier for a camel to go through the eye of a needle than for a rich man to enter the kingdom of God.' It is not just the Christian religion that condemns it though. Almost all religions consider greed as immoral and evil. Judaism teaches that it obstructs other people's opportunities to get what they deserve, Islam mandates charity and generosity to defend against it, and both Buddhism and Hinduism view it as a poison that generates bad karma or prevents spiritual development.[5]

The evolutionary case for greed

Greed is a core aspect of human nature. Like many psychological traits, it is normally distributed within the population, meaning that the majority have intermediate greed, while a few are very minimally or excessively greedy.[6] This individual tendency to greed, measurable using a variety of psychological assessments, tends to remain stable within a person over time,[7] suggesting that it is largely hard-wired within our psyches, less susceptible to the vagaries of our experiences. And as with any such widespread human trait, there must be some sort of evolutionary directive for us to exhibit greed, not just to support the principles that underpin our modern societies. There must be some benefits that propagate greed through the generations.

Therefore, before considering greed as a sin, let us contemplate the good of greed. At first glance, this might seem rather obvious. It is, however, not as simple as just imagining that greed drives people to do better. The acquisition of wealth or possessions is not always inevitably a good thing for the individual. There is a difference between wanting the best outcome, and wanting more. Maximising the outcome of any situation relies upon the balancing of costs and benefits, a rational process to achieve the best, and how you define that. A greedy person, though, may reap negative consequences of their greed, going into debt for example to buy more, or alienating those around them in their relentless pursuit of wealth. The constant dissatisfaction and need for more may cause irrationality.

If we do indeed hypothesise that greed is under evolutionary influence, what might be the mechanisms driving it? Does greed lead to more procreation and increase your chances of passing on your genes?

Unravelling this is more complicated than it first sounds. Greed may manifest in a drive to want more children, but it may also push towards having as many sexual partners as possible, at the expense of forming stable, long-lasting relationships and having children. Greed may drive people to invest more in their social relationships to achieve sexual encounters, albeit temporarily, before moving on.

In the wonderfully titled research paper, 'Greedy bastards: Testing the relationship between wanting more and unethical behavior', being disposed to greed was significantly associated with acceptability and self-reported engagement in unethical behaviours. These included evading fares on public transport, illegally downloading movies, switching price tags at the supermarket and spreading gossip.[8] (The same study showed that being greedy was also associated with increased likelihood of accepting a bribe in a laboratory-based game.) Importantly, this study also found a strong association between greed, the desire to cheat on a partner, and actually doing so.

But does this predisposition to infidelity result in more children, fewer, or make no difference whatsoever? In the real world, or at least the modern European version of it, greed does not appear to correlate with more offspring. In one study, of 2,367 individuals representative of the Dutch population, greed was actually associated with fewer children but, as expected, with higher numbers of sexual partners, and shorter relationship lengths.[9]

Of course, in our evolutionary journey, the modern age with its modern sensibilities is but a blink of an eye. As with those genes that influence obesity, the advance of genes promoting greed may relate to circumstances different to the current era. Perhaps this association between greed and multiple sexual partners found in the Dutch study might be of more relevance to the evolutionary pressures favouring greed, in our past rather than present. Maybe, in different cultures or times, greed might give rise to more rather than fewer offspring. Particularly when having children was less associated with self-sacrifice (for males, at least), having access to more resources would probably have been an important driver of having more children and their increased likelihood of survival. Furthermore, pursuing a strategy of multiple brief sexual encounters might have been a successful evolutionary strategy to have lots of offspring who are genetically diverse.

Greed, money in your pocket and happiness

In contrast to these evolutionary aspects, greed's economic benefits are actually very clear. In laboratory tasks, greed has been
shown to be associated with trying to use time productively, to
reach goals and make progress,[10] to work harder and earn more
money.[11] In the real world, outside the laboratory, greed has
been found to promote performance, mediated by the need for
social status.[12] Of course, there is also the potential that too
much greed might hinder your earning power, since working
well with others, sometimes reining in self-interest, is important
for some higher-earning jobs.

Money is an important motivator for most of us. Even the
thought of money can dramatically alter our behaviour. In a
remarkable series of studies,[13] the effects of mentally implanting
the concept of money into a range of tasks was made obvious.
In one experiment, subjects were brought into the laboratory
and were given a list of sentences that had been jumbled.[14] Some
were given sentences that had words such as 'wealthy', 'gift',
'cash' and 'silver' scattered within them. Others had words that
were more neutral, that had no relationship to money. The
subjects were asked to make sentences out of the list they were
provided. After this task, they were given a problem to solve,
one that required a flash of insight. The researchers found that
those subjects subconsciously primed for thinking about money
were much more likely to persist in the task without asking for
help than the non-money group.

In further studies, participants were again subconsciously
primed to think about money, this time by being asked to either
count a stack of cash notes, or a stack of money-shaped paper,
overtly as a measure of their physical dexterity.[15] These individuals were subjected to physical pain, by plunging their hand
into a bucket of hot water, or social pain, through a social

interaction task where they were ostracised. The money-primed group felt both less physical pain and less hurt in the face of being socially excluded.

In conjunction, these studies imply that the very concept of money tends to make people more self-sufficient, to ask for help less, to rely on their own abilities, but also make them feel more capable, more resilient and to feel less need for others. Money operates as a resource that confers a feeling of being able to cope, to satisfy one's own needs.

The concept of money has another important effect. Psychologists sometimes measure our ability to see the world from our own and other people's perspectives using an almost implausibly simple technique. If I ask you to trace a letter E on your forehead, you can do it in two ways. You may trace a letter E from your viewpoint, or a letter E from other people's viewpoint. Unbelievably, this task correlates well your centre of focus, whether you prioritise your own perspective or those of others.[16] If you ask people to replace the task with an S rather than an E, but ask some to draw a dollar sign rather than an S, those drawing a dollar sign will be more likely to draw it from their own perspective.[17] Money-priming makes you less likely to view the world from other peoples' viewpoints. But evidence suggests that this is not purely a case of money making you more selfish. It appears that money promotes the self-sufficiency mindset, making people less likely to help, but also less likely to ask for help.

Wealth is certainly linked to happiness in some respects. However, absolute income in itself seems to have no bearing on happiness beyond a certain low level. There is little relationship between gross domestic product of a country and average happiness, either when comparing countries to each other or when looking at a specific country over time. In contrast, though, there is robust evidence that on an individual level, relative wealth compared to those around you is clearly correlated with subjective feelings of well-being or happiness.[18]

Wealth, greed and happiness do not necessarily align, however. While greed is associated with a higher household income, it does not equate with a higher personal income (except for certain occupations where the individual is self-employed), and is associated with lower well-being.[19] There are several potential explanations for this disparity between personal and household income: greedy people may induce their partners to work harder; the greedy may select partners who are economically better off; or that the association between greed and fewer children might result in both partners, rather than one, being able to work more. The net result is the same: that the partners of the greedy contribute more to the overall income of the household, despite the greedy not necessarily earning more themselves.

Several other studies also show a relationship between high levels of greed and poorer psychological outcomes. The relationship between greed and lower life satisfaction is consistent, perhaps due to constant dissatisfaction being the central characteristic of greed, or that greed affects the formation of stable, satisfying relationships.[20] Having good relationships is fundamental to well-being, more so than income.

Greed may not just inhibit the formation of stable long-term relationships because that is what the greedy are looking for. Greed hinders so-called prosocial behaviours: sharing, donating, comforting, volunteering and other acts that benefit society. The greedy may have diminished empathic concern – the ability to sympathise with other people – resulting in an impairment of prosocial acts.[21] This in itself is not conducive to sustaining good relationships. There are echoes of Ebenezer Scrooge in Dickens's *A Christmas Carol*, trembling in the presence of the Ghost of Christmas Yet to Come –

embittered, lonely and facing an empty life. Greed is associated with people being more depressed, more anxious, more unhappy and more aggressive.[22]

It is not all about me

Beyond the consequences directly for the greedy, however, there are other reasons why greed has been viewed so negatively by religions and societies alike. The case for greed being a destructive psychological trait to those around the avaricious is very strong indeed. It encourages waste, and the accumulation of unnecessary resources in the hands of one, and restriction of those resources for others.

As we have already seen, greed is strongly associated with the acceptability or engagement of behaviours that break moral codes of fairness, like bullying in schools, running a red light in a car, cheating in an exam or on tax returns, infidelity and accepting bribes. All this unethical behaviour harms others. Greed has been implicated in financial scandal after scandal, like Enron, Bernie Madoff, Jordan Belfort ('the Wolf of Wall Street'), and the 2007–8 financial crisis. Despite the loss of somewhere between \$11–19 trillion in the recession at that time, driven by excessive risk-taking and greed, virtually no top executives of the large banks or other financial institutions went to jail.[23] Like the story of the Panama Papers, this highlights the passive acceptance that greed is a justifiable component of how our systems function.

Greed causes people to focus only on their own needs and wants, at the expense of concepts of values of right, justice and norms, driving activities such as deception, fraud and theft.[24] Psychological studies correlate greed with these sorts of bad behaviour in experimental games in the laboratory, not just as associations in the real world.[25]

Furthermore, in keeping with the view that greed diminishes prosocial behaviour and makes for difficult relationships, this trait is associated with other psychological tendencies, none of them good. Neuroticism, lower self-esteem, less trust in others. Emotional instability, envy. Even psychopathy, narcissism, antagonism and Machiavellianism.[26] These personality traits do not make for a good partner or friend.

The neural basis of greed

In the absence of cases of neurological disorders causing greed, as we have seen with the other sins, there are no pointers as to where in the brain greed originates, at least when it comes to disease. There is no one in the literature whose stroke or tumour has suddenly resulted in isolated avarice, to help identify the greed centre or centres of the brain.

Overtly, compulsive hoarding – excessive collecting and saving behaviour – may be conflated with avarice or greed, but its origins are very different. Individuals with this condition will end up with terrible cluttering of their living space, unable to throw anything away.[27] Hoarding is not just about the accrual of material possessions though. It is typically considered a problem associated with obsessive-compulsive disorder (OCD); about one-third of patients with OCD have some hoarding symptoms. In this context, it is anxiety about throwing things away that fuels the hoarding, the thought that something catastrophic might happen with every item thrown in the bin.

In a few people, however, obsessions and compulsions may not be the explanation for hoarding. This morbid accrual of possessions may be an issue of impulse regulation, with links to other disorders of impulse control like pathological gambling, or compulsive shopping. In these individuals, it is thought that the hoarding is driven by the anticipation of pleasure, derived from the acquisition of objects, and an inability to restrain one's urges (like those with sex addiction in Chapter 3), rather than intrusive thoughts or fears of catastrophic consequences at the prospect of throwing something away.

Observations in people with Parkinson's disease would support this view, that some hoarders are driven by impulsive behaviour rather than the fear of throwing anything away. As I described previously, treatment of Parkinson's with drugs that boost or simulate dopamine within the brain can give rise to a range of impulse control disorders. Not just hypersexuality, but other behaviours that stimulate reward such as gambling, shopping and binge-eating. One study found that 12 per cent of patients with Parkinson's in an outpatient clinic exhibited features of excessive hoarding, and that hoarding was strongly associated with other behaviours implying disordered impulse control.[28]

But whether hoarding is obsessive-compulsive in nature, related to the anxiety that ensues at the thought of disposal of anything, or has a basis in impulse control, it does not represent an accurate picture of greed. Arguably, the pathological 'wanting' of stuff, triggered by excessive dopamine, approximates aspects of greed, but this seems rather simplistic. In impulse-related hoarding, the 'buzz' of acquisition is what drives it, like an addiction, whereas greed seems intrinsically more complicated. Greed is tinged by more abstract concepts such as power and status, and ultimately the wider significance of what we crave. Hoarding is the fear of throwing things away, or the hit of a purchase, but does not stray beyond these basic outcomes, and does not require the necessary acceptance that such behaviour may be at the expense of others.

* * *

Perhaps it is not surprising then, in the absence of disorders of the brain characterised by greed, that our understanding of the origins of avarice from within the brain is more primitive than for some of the other sins. The extent of our knowledge comes from laboratory-based experiments, rather than experiments of nature through disease. We do, however, have an inkling as to the basic processes within the brain that underlie it.

Overtly, greedy individuals are more likely to take decisions in the face of personal profit that are high risk to the company they work for or the society in which they live. There are many factors that drive these sorts of risk-taking, but undoubtedly our decisions are informed by our own experiences. When we take a risky decision, we tend to adjust our decision-making according to what we have learnt from previous decisions. Our choices are influenced by whether the outcomes of those previous decisions have been positive or negative. We learn from good or bitter experience. In theory, then, greedy people making risky decisions on an ongoing basis may show differences in how they evaluate the outcomes of their previous choices.

There are tools available to researchers in this field that permit the assessment of outcomes of decisions in the brain. These techniques centre on brainwave signatures recorded by the scalp electrodes on an EEG. If you make a decision, and the outcome is negative, your brainwaves exhibit a particular pattern called feedback-related negativity.[29] If the outcome is positive, in your favour, then the brainwaves exhibit a different pattern, a signal called P3.[30] By monitoring the size of these brainwave markers in individuals making decisions, researchers can assess those brain mechanisms that provide feedback to the choices you make.

In one study, twenty university students studying economics were tasked with a game that characterised financial risk.[31] They were presented with a picture of a balloon on a screen, representing a value of €1,000. Participants had to decide if they wanted to inflate the balloon. With each inflation, the value of the balloon doubled, but if it burst, they would lose all their money. At each level of the game, the risk of bursting with each inflation went up. The students were told that whoever had the most money at the end of the task would take home €100 of real money. In addition to the balloon task, they were all evaluated for greed as a personality trait using a questionnaire.

The researchers found, not entirely unexpectedly, that those students scoring highly for greed were much more likely to inflate the balloon and risk bursting it. Those greedy students, however, also demonstrated that when the balloon did burst, the feedback-related negativity signal, the electrical marker of appreciating a bad outcome, was diminished compared to students low in greed. This implies that non-greedy students appreciate a negative outcome to their decision more, which guides the learning of decision-making, adjusting expectations and future behaviour. In contrast, greedy people have a diminished ability to learn from their mistakes, to change their behaviour according to bad outcomes or punishment.*

It is not just the ability to appreciate a bad outcome that appears to be impaired in greedy people. Their capacity to appreciate a good outcome is also diminished. In a second study by the same group of researchers, participants were asked to complete a resource dilemma task, where players have to make the decision to act selfishly for their own gain at the expense of their partner, or cooperate to benefit all in the game.[33] This time, the subjects were asked to jointly cultivate an imaginary fish-farm with a partner, thought to be another participant but actually a stooge. Subjects had to choose how many fish to take at each round. If both individuals took two fish, this would

---

* This diminished feedback negativity signal, and the insensitivity to negative outcomes, has also been found in psychopaths,[32] who, like the greedy, tend to be more immune to the costs of their decisions to those around them.

maximise overall revenue, if the participant took more than two fish it would increase their revenue at the expense of their partner, and if both took more than two fish, the revenue of both would decrease.

As expected, greedy individuals were more likely to take more than two fish, despite the knowledge that this would be at the expense of their partner, and possibly of themselves too. But the EEG also showed that those greedy participants had a diminished P3, this electrical signature of a good outcome. Greed therefore appears to be closely associated with a lack of sensitivity to both negative and positive feedback, impairing the learning process of decision-making.

Basically, these EEG studies suggest that greed is due to a failure to learn from previous mistakes, or at least to learn more slowly, less efficiently. There are parallels with the greed exhibited by children, the inability in early life to appreciate that collaboration and cooperation, rather than greed, may ultimately be in our self-interest. While a child will have had fewer learning opportunities, a greedy adult will not have learnt as effectively from these opportunities.

It is not just these studies of electrical activity that imply that the ability to weigh up decisions is crucial to being greedy or not. The imaging of greedy brains suggests this too. Functional MRI scans show differences in activity between the brains of greedy and non-greedy individuals, in one particular area of the pre-frontal cortex – the core to so much of our thinking, planning and decision-making processes. This region, the ventromedial pre-frontal cortex (vmPFC), with which we are by now very familiar (see Chapter 1), is thought to play a pivotal role in evaluating rewards and punishments. Indeed, hindrance of this evaluation may also contribute to the other sorts of behaviour we have already looked at, like wrath and lust.

The brains of the greedy are not only different in terms of

activity. There is evidence of structural differences too. The greedy have slightly less grey matter within the vmPFC, as well as other related areas.[34]

When it comes to greed, as with the other sins, the neurological equipment to rein in these behaviours is fundamentally dissimilar. I hesitate to use the word flawed, since as we have seen, greed has positive and negative facets. As with all these aspects of our personalities, greed and its biological origins are a spectrum rather than a binary phenomenon.

The problem with pathologising

We celebrate those who have amassed great wealth, while simultaneously expressing disapproval. We recognise that this character trait has both positive and negative consequences, for the individual and for us all. And therein is the central tenet of our reluctance to pathologise it, to label it a disease or medically abnormal. Within the intricate tapestry of the human condition, the concept of 'greed' remains an elusive thread. For greed is inherently subjective, and judgement of it is coloured by shifting societal norms and values. To pathologise it also requires us to grapple with questions of autonomy and accountability, in our current world where individual choice is sacrosanct.

And because of its subjective nature, even characterising or measuring greed is problematic. Greed is not a monolithic entity, and behaviours associated with it illustrate the complexities of human experience and emotions. Consider those individuals whose wealth dwarfs that of entire nations: according to an Oxfam report in 2020, the richest twenty-two men in the world have more wealth than all the women in Africa; and the world's 2,153 billionaires have more wealth than the 4.6 billion people who constitute 60 per cent of the global population.[35] One might argue that greed – an insatiable desire for more at all costs – is what drives this accumulation of wealth, and the power, privilege and control that come with it. However, I suspect that greed as the ultimate or singular motivation is applicable to only the minority. This 'greedy' behaviour is not just due to intrinsic greediness. It is a mixed palette, tinged by colours of other personality traits.[36] Materialism – the view that the acquisition of possessions is a core aspect of life and crucial to well-being and happiness. Envy. But also, competitiveness, conscientiousness (which often correlates with 'miserliness'), and occasionally a desire to use their wealth for altruistic purposes.

Maybe Gordon Gekko and Saint Paul were both right.

7.

Pride

*'I burn with love for – me! The spark I kindle is the torch
I carry.'*

Narcissus, in Ovid, *Metamorphoses*

As Chris and I discuss his case, there are points at which I strain
to be clear in my own mind as to whether what he is describing
truly represents 'grandiosity' – a heightened sense of superiority
– or indeed, that he was simply right all along. At other times,
I struggle to know where his innate personality ends, and his
brain disorder begins.

'You talk about grandiosity,' he tells me. 'But there is a
grandiosity in each and every one of us, that makes us think
that we are more than we are. Grandiosity is just one of our
evolved strengths, because the stronger male gets the mates.
It is that self-belief, that grandiosity, that is an essential part
of survival in a life-and-death situation. No simp,* no avoca-
do-eating latte-drinker, ever created anything, because it
requires that dominance [to move] forward. You are trying
to say it in a negative context. Grandiosity is a self-belief. I
needed a self-belief to overcome, to live,' he says. 'What has

---

* The term 'simp' is a slang term, referring to someone who shows attention and
  sympathy towards another, usually someone who does not reciprocate the same
  feelings, with a view to engendering affection or a sexual relationship.

happened is that society has become beautified. It is becoming cushion-covered, primrose-smelling softness. Any male trait outside this minority neutering viewpoint is seen as toxic. So, toxic masculinity – another bullshit phrase – is deemed to be wrong. But it is not wrong, it is an essential part [of human life]. It was required for the Industrial Revolution for example, for us to evolve. Let's see how long the lights stay on without it!'[1]

Despite a prematurely curtailed formal education, Chris is clear he has always had that self-belief in his intellectual abilities. He prides himself on his logical mind, his ability to systematically analyse and think through problems in an unemotional, methodical way. On the emotional side, less so. Wendy, Chris's wife of twenty-eight years, says: 'I am an emotional person, but Chris does not deal with emotion whatsoever. If I were to cry, Chris would say: "Why cry? What is that gonna do for you?" So, then I cry more.'

Chris describes himself as always having been different – socially anxious, a little awkward. He says this is the root cause of his lack of formal education. Now in his fifties, the scars of his childhood run deep. 'I was head boy at school. I got nine O levels. But unfortunately, my father wanted me out of the house. I was the odd one out, he wanted me gone. I was always the last in line, despite being the middle one of five. I wasn't dominant enough. A silly example – we used to have a tin bath, five children [bathing] in front of the oven. I used to get the water at the end. Always.'

The army proved to be Chris's escape from the family home at the age of sixteen, although the rigid routine was not for him, and he left before signing on at eighteen. An eclectic career followed: an aircraft technician, an aircraft engineer with the Civil Aviation Authority, an electrician, then construction site supervisor. All the time pursuing his own intellectual interests on the side. 'I'd been academically involved in this ten-year quest to find a solution to something to do with my own hypothesised variant of zero-point energy. Imagine a spectrum of energy. And it's that point at which things almost cease to exist. I theorised that the centre point of a revolving mass has tappable, endless energy,' he explains. 'That is zero-point energy. I wanted to devise a mechanical means to create a physical mechanical object that cyclically transfers force in the fourth dimension, which I did. I got to the proof-of-principle stage before I was hospitalised. So I had to withdraw from my own research and patent-filing process.'

I later look up zero-point energy, to learn that it is the lowest possible energy that a quantum mechanical system may have, with its origin in quantum field theory and broad implications for physics and cosmology. I am quickly bewildered by it.

The nature of Chris's thinking patterns, and his lack of emotionality, have led him and Wendy to consider a diagnosis of autistic spectrum disorder. He has requested a formal assessment. In one of his many emails to me before and after we meet, Chris writes:

As you have no doubt gathered in very short order, I have, in my wholly unqualified opinion, undiagnosed autism, which manifests itself in the compulsive need to enquire and understand through research and pointed, direct communication with others in the field that I am interested in, or pressed to explain my understanding of. Often, that expressive nature of autism is, certainly in my case, met with ridicule by others. It has been said by some in the past, that due to my working-class background, and lack of university education, I simply have no right to use concise, descriptive language when engaging, in my mind, as an equal.*

It was in the winter of early 2015 when Chris first became unwell. In his late forties, he was still working as a site supervisor, overseeing major construction projects, and fully engaged in his zero-point energy project. Christmas had been marred by toothache, an uncomfortable niggle in his jaw. 'I had a tooth abscess in one of my right upper molars,' Chris recalls. 'I just put some Bonjela on it, and tried to carry on through. But then I became more fatigued, and started getting headaches. Jesus Christ, I couldn't return to work at that stage. I would go to work at eight, and by nine, I would be back indoors, asleep on the settee.'

His deteriorating health brought some benefits, however. 'It sounds odd, but the accelerated [mental] processing was absolutely wonderful. I could see the world clearly and diagnose all the problems. In terms of memory recall, it was so acute. I could paraphrase everything that people were saying to me. It was like going through space in an accelerated fashion. The neural links were firing, *bang, bang, bang.*'

Wendy noticed it too. She tells me that he had always had a talent for recalling everything he had read or heard, but in this period, this ability seemed heightened.

As the days progressed though, his health continued to worsen. Wendy recalls the day she insisted he seek medical

---

* When I ask Chris for permission to share his emails, he replies: 'Maybe with a footnote somewhere saying, that all opinions are expressed from the mind of one particular lunatic . . . This insight will be entitled 'The Literal Mechanics of Me', in which, I will detail in a more extrapolated form . . . in my proposed whole new Chapter of Classical Mechanics. Which is how life itself can be illustrated in purely mechanical form. For example, I have sent a derivation of my discovered base geometry to NASA Research, with which a supported payload at the distal end of a passive [object] can spatially moved about in ANY direction, in an effortless manner, as if weightless.'[2]

attention. 'Chris had come in and was laying on the sofa. He just had a strange look about him. Almost jaundiced and pale grey. His left eye had just dropped, and he had some twitching around the eye. So I took him to [the local hospital]. And they said it was probably nothing. But they got a scan, and they saw this black thing. They said: "We think he has a brain infection."'

The scan was transmitted to the nearest neurosurgical unit, but the hospital was becoming increasingly overloaded with patients. While awaiting a response, Chris and Wendy were sent home. 'At nine o'clock that evening, my phone went,' says Wendy. '"Where is Chris? Don't let him move. He definitely has a brain infection." The next thing, the ambulance arrived, and a specialist ambulance car as well. It was absolutely horrific. He had started to make little sense.'

Chris was whisked off back to hospital, before being transferred that night to the neurosurgical unit, a diagnosis of a brain abscess made on the basis of the earlier scan. He recalls feeling very detached, observing everything at a distance, but remembers everything very clearly. Wendy, unable to accompany him in the ambulance, was left at home with their seven-year-old daughter, wondering what was happening: 'It was horrible. I never want to go back to that point ever again.'

The MRI scan from that time shows a spherical abnormality deep in the right temporal lobe, surrounded by massive swelling extending into the rest of the temporal lobe and into the parietal lobe too. The following day, having had huge doses of antibiotics delivered via a catheter (a long thin tube inserted in the arm passed through to the larger veins near his heart), he was taken to surgery. Surgeons drilled a small burr-hole in his skull before draining the abscess. Upon analysis of the pus, his antibiotic regimen was switched to a more targeted treatment. He was also prescribed high dose steroids to reduce the swelling in his brain.

Only mildly religious before his illness, Chris has a vivid memory of a vision as he was being anaesthetised for his operation:

I found myself on the dock on this mist-covered lake. And this boat came in with the captain, the ferryman. I intuitively knew what he was or what it symbolised. And I looked and saw loads of people in his boat, so sad, heads down. And I said: 'I don't want to go, I do not want to go.' He said to me: 'I've seen enough sadness today to last a lifetime. I'll give you one chance by asking you a question. If you get that question right, you can go, but if you get it wrong, you're gonna have to get on the boat. So he asked me the question. He goes: 'Okay. Do you accept the terms?' 'Yes, I do.' 'What is life?' I said: 'Life is no more than time. Time to love, time to experience, time to cry, time to enjoy. That is what life is no more than that. At the end of your time, when you are literally in your time, what would you do or pay to get back all the wasted seconds, minutes, hours, days on people who didn't appreciate your time? It is the most valuable asset known to man.' And he said: 'Yeah, you're right. What do you want in return?' I said: 'I want fifty years with my wife,' and he said: 'You got it.' I said: 'What do you want me to do with my time?' 'It's your time. Do with it as you will.' This is a very distinct memory.[3]

When Chris regained consciousness, Wendy was holding his hand.

Wendy's experiences of that day were rather less spiritual. In the middle of the night, Chris's sister, living in Southeast Asia, had phoned the hospital to find out how her brother was, only to be told in error that he had died. When Wendy called the ward, they were unable to tell her where Chris was, and she jumped in the car at two in the morning and drove the seventy-five miles to Chris's bedside, in dread all the way.

The abnormality on the brain scan did not shrink quite as fast as expected. Another operation to remove debris, and five weeks on the neurosurgical ward receiving more antibiotics and steroids, followed. A condition of discharge was major dental treatment. The tooth abscess, presumed to be the source of infection that had seeded to his brain, was discovered to be far more extensive than previously appreciated. Chris had all his teeth pulled out, every tooth socket full of pus. He flashes me a tooth-baring smile and taps his immaculate front teeth. 'Every single one, false,' he laughs.

Physically, he had made a partial recovery. 'I was still physically strong, but was not able to connect my brain to my legs. I needed support.' It was quite clear, however, that behaviourally not all was well either. Wendy describes him as being very self-focused, at the expense of everyone else. She tells me the experience of their daughter, who found his inability to engage with her very distressing.

Chris describes this aspect of his thinking as being 'necessary', crucial to his healing. He strongly felt that he was not getting the help that he required to get better. 'I needed to devise a plan of recovery. It sounds arrogant, but I had a Plan B – to use logical analysis and critical thinking, to simplify my situation, to extract all emotion from it.' Chris was convinced that he would have to orchestrate his own return to normality, and that this could only be achieved through single-minded self-care. 'In extreme circumstances, you have to love yourself first. A point of survival. You have to, because what good are you to your family if you don't try to get back to where you were?'

Wendy tells a different story, recalling the train journey home as 'absolutely horrendous'. On picking Chris up from the hospital, she describes a totally different man to the one she had known before. 'He was quite arrogant. He was quite abusive, swore at everyone on the train, [Chris] trying to smoke on the train.' Chris remembers this well. 'I had an overwhelming sense of self-belief, of self-worth. A feeling of superiority. It came out as arrogance. I thought I was God's gift. I had an accentuated sense of how handsome I was.' Wendy says this sense of superiority also manifested in Chris's manner to her. He became belittling, putting her down. 'I was his carer, not his wife. Everything I did was wrong. Everything I touched, everything I said, he would say was just so wrong.' The journey home, managing Chris, trying to rein in his behaviour, was incredibly stressful. But things were about to get much, much worse.

They wrongly thought Chris's medical issue, the brain abscess, had all been dealt with. 'We were on the train, and I got a phone call. "Oh hi, is that Wendy? This is [a nurse] from oncology. Is Chris still in hospital?" And I said no, and she went: "Okay, don't worry, I'll call you some other time." And I was like: "No, no, no. Stop!" because this was the first I'd heard of oncology. We had been given the discharge summary,' she continues, detailing the diagnosis of a brain abscess, the antibiotics, the steroids. '"Hang on a minute. You've just said oncology. The whole time Chris has been in the hospital, I've not heard the word oncology. What do you want?" And she said: "He has grade 4 GBM.* We need to get you referred to local oncology."' Wendy sank to the floor in the middle of the train carriage on all fours, sobbing: 'It was honestly the worst day of my life.'

* * *

---

*   GBM, or glioblastoma multiforme, is the most aggressive form of brain cancer, with a very poor prognosis, and a median survival of a few months. All GBMs, by definition, are grade 4, according to the World Health Organization grading system for classifying the aggression of tumours.

They fuck you up, your mum and dad.
They may not mean to, but they do . . .[4]

Parenting these days is very different. For most of us growing up in the 1980s or earlier, childhood was very laissez-faire. Our parents fed and clothed us, checked that we were safe, and if we were lucky, engaged with us when time permitted. These days, we are not doing our parental duty unless we are holding our child's hand at every step of their journey. It is widely considered that the success and happiness of our children is dependent on their self-esteem, and that we as parents play a crucial part in building it.

Children of this new way of parenting seem to benefit. I see it in the medical students I teach: impressive young people, many of whom I am in awe of, carrying themselves with a degree of self-possession, self-confidence and fearlessness that my generation would never have dreamed of. These young people would not for a second tolerate the teaching methods of the past, the ritual humiliation and terror of an old-school medical education, of never speaking up unless asked, of not daring to interrupt or question the consultant. We would never have challenged an exam result, complained about the quality of teaching, or the belittling by our seniors. This self-esteem, a sense of one's worth as a person, is typified by an interest in self-improvement, of personal growth. There is a robustness of character in the face of adversity, an ability to cope with failure or poor performance, to pick themselves up and keep going. And a realistic view of their own ability, recognition of when they have done particularly well or especially badly.

Every so often, however, there is the medical student who is frankly deluded. Despite consistently failing, they cannot accept that their performance is poor. They believe they are clever, talented and able, despite a wealth of evidence to the contrary. They are boastful, overconfident and convinced of their intrinsic superiority over their peers. And in the face of negative feedback, they are unaccepting, simply unable to consider the accuracy of a poor evaluation. You will undoubtedly have met such people at school or in the workplace. You might be forgiven for thinking that these individuals have extremely high self-esteem. But they are actually something quite different. They are narcissists.

In contrast to self-esteem, narcissism is defined as an inflated sense of one's own importance or merit.[5] Like greed, it is a psychological trait, and all of us sit somewhere on a scale of narcissism. At the extreme of the scale, individuals with high levels of narcissism will fantasise about their degree of personal success, will believe that they are special or superior to others, deserving of respect and admiration, and should be treated differently to others.[6] When humiliated, they are prone to lashing out, verbally or even physically. These qualities are not those associated with high self-esteem, and indeed self-esteem and narcissism are only weakly correlated.[7]

While high self-esteem is associated with realistic evaluations of oneself, narcissism is characterised by an illusory self-view. Narcissists will consider that they have performed extremely well, despite failing at a task. Rather than being driven by a desire to improve oneself, a narcissist will be focused on getting ahead of others, demonstrating their superiority, showing little concern or empathy for those considered their inferiors. And in contrast to those with high self-esteem, who exhibit robustness in the face of failure, narcissists will feel great shame, and may respond aggressively to negative feedback.[8] Over time, shame may give rise to anxiety and depression, both of which are associated with high narcissistic traits.

We tend to think of most psychological traits as stable. Greed, wrath and envy, or indeed kindness, generosity and empathy, are presumably as common now among us humans as they were 2,000 years ago. But this may not be the case for narcissism. It has been suggested that between 1979 and 2006, the prevalence of narcissism in college students rose by about 30 per cent.*[10] Furthermore, rates of narcissism are higher in the US, Canada and Europe, where societies tend to be more individualistic, than in Asia or the Middle East.[11] This implies that a relatively recent change in our environment has influenced a shift in levels of narcissism, something culturally specific. And as Philip Larkin suggests, in his poem, *This Be The Verse*, our parents bear a great deal of responsibility. The finger of blame has been firmly pointed at our parenting style.

Two differing theories are at play when it comes to the origins of narcissism. The first, termed social learning theory, proposes that children are likely to become more narcissistic if their parents overvalue them, thinking that they are more deserving or special than other children. If children are told they are special, they grow up believing they are special. It is difficult to prove this theory, however. If you ask a narcissist whether they were highly valued and praised by their parents, the answer is going to be predictable. Narcissists often remember their parents as overvaluing them, but this is perhaps not that surprising. Since narcissists often feel admired by others, they might be expected to remember their parents as being admiring of them too.

In contrast, psychoanalytic theory proposes that it is lack of parental warmth that drives narcissism. A lack of attention, appreciation and affection from their parents pushes children to elevate their sense of self-importance to gain approval from others, in the absence of receiving it from their parents.

To clarify which of these theories might be correct, a Dutch group of psychologists studied a cohort of 565 children aged between 7 and 12.[12] In this age group, narcissistic traits can be

---

* This figure is debated. Some authors have criticised the methodology surrounding this study, and have highlighted the wide-ranging implications of labelling generations as more narcissistic.[9]

properly assessed, and individual differences in narcissism
emerge. Every six months, the children and their parents filled
in a range of questionnaires regarding levels of narcissism,
self-esteem, parental warmth and parental overvaluation.
Parental overvaluation, both from the mother and father,
predicted narcissism in the child over time, but did not influence
the child's self-esteem. In contrast, lack of parental warmth had
no influence on narcissism, but child-reported parental warmth
did negatively predict self-esteem.

Basically, the researchers concluded that parental overvaluation
drives narcissism, while perceived lack of parental warmth fuels
low self-esteem. There is an alternative explanation for these findings, though. Perhaps narcissistic parents overvalue their offspring,
because they see them as an extension of themselves. Hence, children of narcissistic parents might be overvalued, but the root cause
of the child's narcissism is either genetic or learnt through copying
their parents. The researchers, however, found relatively weak correlation between parental narcissism and parental overvaluation.
They concluded that the origins of narcissism in children is in
keeping with social learning theory – 'children come to see themselves as they believe to be seen by significant others, as if they
learn to see themselves through others' eyes.' Self-esteem, in
contrast, appears to be related to parental warmth – 'When
children are treated by their parents with affection and appreciation, they may internalise the view that they are valuable
individuals, a view that is at the core of self-esteem.'[13]

It has been proposed that in our quest as parents to instil self-
esteem in our children, we have been going about it the wrong
way. By telling our children how special they are, not holding
back on praise, treating them as exceptional human beings,
we might be pushing them towards narcissism rather than self-
esteem. Instead, we should be treating them with love and affection, while encouraging realism through feedback, and avoiding

excessive praise. We should be fostering self-improvement rather than cultivating a drive for superiority. We should avoid engendering fragility by ensuring that they do not feel that their worth is dependent on meeting our own standards, by accepting them for who they are, even if they fail.[14]

Chris's description of his upbringing is certainly indicative of a lack of parental warmth, but also of an under-evaluation, rather than being placed on a pedestal by his parents. These might be expected to drive both low self-esteem and low levels of narcissism. His childhood is, therefore, from my perspective, an inadequate explanation for his latent heightened self-belief, or for his grandiosity after discharge from the hospital.

The shifting meaning of pride

It is not just our parenting styles that have changed in the last few decades. It is also the meaning of pride. If I hear the word now, for me it has almost entirely positive connotations. Self-worth, self-esteem, confidence, accomplishment, dignity – these are hardly sinful outcomes. The Pride of the LGBT+ and the disability communities, celebrating difference and diversity, opposing shame and judgement. A call to self-worth and self-esteem. The pride in our own ethnicity or culture, particularly if we are in the minority. Pride in our family, our work, our beliefs.

It seems incomprehensible, through modern eyes, that pride not only be considered a sin, but be elevated to the most serious, decreed by Pope Gregory as the root of all evil and 'the queen of sin', in 590 CE. This paradox highlights what the Dutch researchers above found, that there are two psychological traits, only vaguely related: those of self-esteem and narcissism, sometimes termed as 'authentic pride' and 'hubristic pride' respectively.

Indeed, such a distinction has long been recognised. Aristotle thought of authentic pride as the pinnacle of the virtues, but distinguished this from the vice of hubris, in terms very familiar to today's readers:

> For he who is worthy of little and thinks himself worthy of little is temperate, but not proud; for pride implies greatness . . . Pride, then, seems to be a sort of crown of the virtues; for it makes them more powerful, and it is not found without them. Therefore it is hard to be truly proud; for it is impossible without nobility and goodness of character . . . Hubris is not the requital of past injuries; this is revenge. As for the pleasure in hubris, its cause is this: naive men think that by ill-treating others they make their own superiority the greater.[15]

Authentic pride can have positive outcomes. While pride may follow success, evidence points to pride also driving success. Pride is not only a behaviour, but also an emotion. That feeling of having done something well, the pleasure or exhilaration of achievement.

And it seems that this emotional force changes how we think about situations, making us feel that our achievements are under our own control, hence pushing us to persevere.[16] Instilling pride through praise, regardless of true performance, can induce harder work and perseverance in tasks.

That pride has this effect on us can be seen in the psychology lab. One experiment involved 109 psychology undergraduates, who were told they would be interviewed regarding a number of college-related topics.[17] After a few minutes of the interview, their emotions were manipulated, by inducing anger, shame or pride. The students were randomly assigned to one particular emotion. A stooge came in and gave feedback: hostile demeaning statements like: 'Are you kidding?' or 'Where did you get that idea?' to trigger anger; statements like: 'I don't think you are doing as well as you can, can you try a little harder?' and: 'I think you can do better on the next issue', to induce shame; and praising remarks like: 'You are doing a lot better than the other students', and: 'You're doing great; you have brought up a lot of interesting points', to induce pride. The task then continued.

At the end of the task, independent judges watched the videoed interviews and evaluated how well each individual student had performed. Those students who had been primed for pride performed better, and reported that the task was lower in threat, lower in demand, and fairer than the other groups. Furthermore, the pride students reported feeling a higher degree of control over the proceedings, implying a greater degree of perception that their own actions influenced their success. And if you feel that your own actions correlate directly to success, then you might be inclined to persevere a little longer at any particular task.*

Hubris

There is a large grey area, however, between this sort of authentic positive pride, with its benefits, and the pride of hubris or narcissism. A quick glance through social media illustrates this well. Countless tweets or posts starting with: 'I am proud to have been asked/awarded/mentioned . . .' (increasingly, the word 'proud' is being replaced by 'humbled', although to all intents and purposes, the meaning is the same) can generously be interpreted as authentic pride. Then there are others, boastful and arrogant; the racist, homophobic, antisemitic, dripping

---

* A similar study showed that pride manipulation induced students to continue at a tedious task involving mental rotation of three-dimensional objects for about 40 per cent longer.[18]

with misplaced superiority.* The hubristic messages of some of our politicians. All of these clearly cross the line into narcissism, into hubristic pride. Politicians in particular seem to suffer from hubris – 'impetuosity, a refusal to listen or take advice and a particular form of incompetence when impulsivity, recklessness and frequent inattention to detail predominate'.[20] These are the words of the politician and physician David Owen,† arguing in 2009 for 'Hubris syndrome' to be formalised as a psychiatric disorder. He defines it as an exaggerated pride or narcissism that he characterises as akin to a personality disorder, albeit coming on in later life, after leaders have been in power for many years (unlike personality disorders that are usually evident by early adulthood).

Owen, and his co-author Jonathan Davidson, identified seven US presidents and seven UK prime ministers exhibiting features of this syndrome – namely a narcissistic tendency to see the world as an arena in which to exert power and seek glory; to act to portray themselves in a good light; a disproportionate concern with image and presentation; a messianic demeanour;

---

* The association between social media use and narcissism has garnered some attention in academia. There are distinct features of communication through social networking sites that are of particular attraction to people with narcissistic tendencies: the opportunity to broadcast self-related information to a large audience, and get positive feedback through likes; the selectivity of what information can be revealed; the use of posts or photos to embellish success and superiority, while news that does not conform to a narcissist's own view of themselves can be omitted. Certain patterns of behaviour on these sites, especially uploading lots of photos of oneself, and the number of friends or followers, correlates with grandiose narcissism.[19] The strength of this correlation varies between countries. In cultures where social stratification is high, and the status of citizens is viewed as firmly fixed, this link between narcissism and social media behaviour is stronger. Presumably, social media engagement is a rare method to express uniqueness or specialness in these sorts of societies, and therefore is even more attractive to narcissists.

† Prior to his political career as a Labour MP, and subsequent membership of the 'Gang of Four', who broke away to form the Social Democratic Party, Owen was neurology and psychiatry registrar at St Thomas' Hospital in London, where I currently work.

a tendency to speak about oneself in the third person or using the royal 'we'; exaggerated self-belief and self-confidence to the extent of showing contempt for the advice or criticism of others; a belief that they are accountable only to a higher power such as History or God, rather than their colleagues or the public; and a loss of contact with reality. Since the publication of that paper over a decade ago, there have of course been other candidates for Hubris syndrome in both these offices of state.[21] I will leave it to you to decide who fulfils these criteria.

For Hubris syndrome, prolonged power, particularly when associated with overwhelming success and limited constraints, is proposed by Owen to be the trigger. Our leaders are probably enriched in certain attributes that prime them for hubris anyway. A degree of impulsivity is a prerequisite to making big decisions on limited evidence, and having the hide of a rhinoceros, being immune to criticism, is crucial as a politician. As Owen and Davidson write: 'They [these attributes] make up the pores of the filter through which such individuals must pass to achieve high office.'

The personality of pride

Hubris syndrome is proposed as leading to some political decisions of global importance, resulting in outcomes such as the Watergate scandal, the 2003 invasion of Iraq and the appeasement of Hitler in 1938.[22] However, the underlying tendency to hubristic pride or narcissism is not just limited to politicians and the world stage.

As I have already described, we all sit somewhere on a continuum of narcissism as a psychological trait. To have a slightly elevated sense of self-importance can be a useful character trait in some walks of life. For most narcissistic people, like those rare medical students who remain oblivious to their abilities, it is more an irritation to others rather than a destructive force. For some people, though, their narcissism is so extreme that it impacts very negatively on the individual and those around them. It is at this point that an exaggerated feature of one's personality becomes a disorder.

We have already touched upon the two faces of narcissistic personality disorder (NPD) in the context of envy. NPD is not rare, with lifetime prevalence estimates as high as 6.2 per cent (slightly higher in men than women).[23] While NPD is typically marked by grandiosity, arrogance, dominance, the pursuit of power and a general disregard for others, some individuals exhibit a profound vulnerability to criticism, low self-esteem, and deep shame. It is this latter group who often suffer from severe envy, and are at high risk of anxiety, depression and suicidality.[24]

Overtly, these two aspects of the same condition – on the one hand an inflated self-importance, on the other a fragility of character – are strikingly different. Consider these different examples of individuals with NPD, presented in a paper:[25]

The first is a man in his forties, a successful entrepreneur, competitive, and highly sociable; the 'life and soul of a party'. His personal life, however, does not match his professional life. Since early on in his marriage, he has lost any sexual interest in his wife, and he has had serial affairs, financially supporting his lovers before getting bored, and suddenly cutting them off. He believes his affairs have no impact at all on his relationship with his wife, but wonders if he could do better, and considers leaving her.

The second is an unemployed man in his mid-thirties, with a background of cocaine and alcohol abuse. He is seen in the emergency department claiming that he has had a dental procedure and requests opiates. Initially extraordinarily ingratiating, when the doctor tells him she will need to speak to the dental surgeon who performed the procedure, he turns, and rapidly becomes insulting and bullying. When the doctor phones the woman listed as his next of kin – his girlfriend – she tells the doctor that she is no longer the man's girlfriend, as he had been exploiting her financially. He has been fired from his job in the financial sector, and cannot find another job that meets his expectations. Instead, he has been living off her and off his father.

The third is of a man with diabetes, requiring insulin. He has had a series of low-level jobs but has never really found his feet in the workplace. His mood is low, and when it dips, he might 'forget' to take his insulin and end up in hospital due to loss of control of his blood sugar levels. He constantly compares himself to others, describing himself as lacking in some way, but simultaneously feels slighted that his talents are not appreciated. He fantasises that his employer recognises his abilities and promotes him, while at other times also has fantasies of humiliating his boss through his superior knowledge.

At face value, these individuals have little in common, but they all have narcissism underlying their situations. The second case demonstrates grandiosity and arrogance, with bullying or coercive behaviour; the third is a good example of vulnerable narcissism, where more covert feelings of grandiosity are intermixed with shame, negative self-views and a fragility in the face of criticism from others. The first case is of someone who in many aspects of their life uses their narcissism to their advantage, achieving success through their exaggerated personality trait, sometimes termed 'high-functioning'. Nevertheless, NPD clearly influences this man's ability to maintain healthy long-term relationships.

At their core, all these individuals have two shared features that unify them. The first is that, for most people, their view of their own identity is informed by experience, and indeed reality. For people with NPD, however, their sense of self is fragile, and less informed by reality. Their view of themselves is heavily dependent on a self-perception that they are special or exceptional, not like everyone else. Maintaining that view can be problematic, as it requires a denial of aspects of reality that contradict the view that they are amazing or touched by the hand of God. Unfortunately, engagement with others on a deeper level is often an opportunity to threaten that own self-view. Grandiose narcissists will therefore enter into superficial or coercive relationships that act to boost their grandiose sense of self, while vulnerable narcissists tend to withdraw from relationships more fully, so that no evidence challenging their sense of superiority has the opportunity to be presented.[26] The outcome of narcissism may be influenced by other personality traits. Grandiose narcissism is strongly linked to being an extrovert, while vulnerable narcissism is associated with neuroticism (traits that predispose to negative emotions such as anger, anxiety, irritability and depression) and introversion.[27]

Therefore, the preservation of a sense of identity that is inflated in its import or ability leads people with NPD to have little interest in the *genuine* views or feelings of those around them. They want others to see them in the same light they see themselves. In NPD, the default is to project one's own views of oneself onto others rather than understand the reality of any interpersonal relationship. Understanding the feelings in someone else's mind is essentially the definition of emotional empathy.

And here is the second unifying feature of NPD: people with NPD exhibit clear deficits in certain aspects of empathy.[28] People with NPD often demonstrate extremely good cognitive empathy – the capability to figure out someone else's emotions and motivations. This skill is really rather important when it comes to knowing what makes someone tick, and manipulating them for one's own purposes. In contrast, people with NPD show little emotional empathy – *feeling* the emotions of others, rather than simply *knowing* them. For grandiose narcissists, the lack of emotional empathy is based upon disregarding the feelings of others, while in vulnerable narcissists, the intensity of self-consciousness or worry about themselves simply overrides other people's perspectives.

Traces of this impairment of emotional empathy can be seen on brain scans in people with high levels of narcissism. Functional imaging shows that specific regions of the brain are activated during the experience of emotional empathy in non-narcissistic individuals. These sorts of brain scans look somewhat different in narcissistic individuals. One study took subjects with high and low levels of narcissism and, while scanning their brains, showed them images of faces expressing emotions. The biggest differences were in those very areas implicated in emotional empathy.[*29] These same regions are also found to be different in volume, not only in activity, when comparing normal controls and patients with NPD.[30] The neurological apparatus necessary for emotional empathy is smaller in those with narcissism.

Pride beyond personality

While excessive displays of 'pride' or enhanced self-importance are part and parcel of the spectrum of human experience (albeit sometimes elevated to the level of a personality disorder), there are more dramatic examples of pride – as in Chris and Wendy's

---

[*] These brain regions include areas considered important to the integration of information from the external world and our inner selves. These parts of the brain have been proposed as acting as a switch between the central executive network, involving the pre-frontal regions and allowing an individual to behave in a way appropriate to the external context, and the default mode network, circuitry related to self-reflection or inner thoughts.

train journey home from the hospital. Sometimes, pride reaches a feverish intensity that meets the definition of delusions, those false beliefs about external reality that are unshakeable, despite overwhelming evidence to the contrary. Delusions, alongside hallucinations, are 'the basic characteristics of madness',[31] wrote the Swiss-German psychiatrist and philosopher Karl Jaspers. More precisely, the hallmarks of psychosis. People with psychosis experience a very wide range of delusions – from beliefs about being able to broadcast their thoughts or read minds, to persecutory delusions.

Sometimes delusional beliefs can be rambling, broad-ranging and disorganised, but they can also be highly specific. Such a laser-like focus and clarity of content in delusions is remarkable when encountered. I recall one young woman, working in an extremely demanding job, who walked into my clinic room every inch the professional – polished, groomed, dressed immaculately in an expensive suit, and clutching a designer handbag. I assumed she had come to consult about her migraines or her sleep, but was dumbfounded when she began to tell me the reason for her visit. She was totally and utterly convinced that the family of a suitor she had rejected had implanted a device in her brain that permitted them to monitor her thoughts, and even insert their voices into her head. No number of X-rays or brain scans could persuade her otherwise. In intricate detail, she explained the technology of this device, how it worked and where it had been inserted. In all other respects, she appeared totally normal.

In the context of pride, however, there is one class of delusional thinking which is particularly relevant: that of grandiose delusions (sometimes referred to as delusions of grandeur) – the unfounded belief that you have special powers, wealth, identity or mission.

Grandiose delusions are very common in patients with psychosis, occurring in roughly half of people with schizophrenia and two-thirds of those with bipolar disorder. They are also well described, both in the medical literature and in popular culture: Stan Laurel dressed as Napoleon in the film *Mixed Nuts* (1922) and being dragged off to a lunatic asylum is one of many that comes to mind. Despite their frequency, these sorts of delusions are poorly understood, perhaps in part as they are perceived as being less malign than other types; to imagine you are Jesus Christ seems intrinsically less threatening or dangerous than believing that everyone is out to get you. They are also considered a symptom of an underlying disorder, and so the focus for research has been on the root cause rather than its manifestation.

However, this relatively benign view of grandiose delusions is not entirely correct, as interviews with patients illustrate.[32] There are, of course, the physical dangers of an intense belief that you can walk on water or fly. 'In some cases I wouldn't think through where I tried [walking on water],' commented one patient. 'So maybe it will incidentally be shallow . . . but also in deeper places . . .where getting out might have been challenges. I could've got seriously hurt.' On trying to fly, said another, '[I] stepped off things and expected to fly.' Or believing you are invincible and stepping out into moving traffic.

There is also sexual harm, where the grandiose delusions have placed people at risk of sexual exploitation 'This elderly gentleman came up to me . . . I thought: "You're God." I went to his house . . . We had some kisses and cuddles and I said: "Can we be married?" He said: "No . . . we can be partners,' and from that I thought he meant not literally romantic partners but business partners; partners in the process of saving people,' reported one of the patients interviewed in this paper. And there is also the distress or anger of being rejected or isolated through these delusions, or the pressure, for example, that comes with being the Messiah and having to save mankind.

Delusions of grandeur can also have a positive aspect too, however. Many grandiose beliefs centre on special powers or roles that help society, such as being Jesus, or working undercover for the security services. These beliefs bring with them a meaning to life, granting a sense of purpose or self-identity. They often have a flavour of a protective mechanism, occurring in the context of negative circumstances, providing relief or making sense of suffering.[33]

These life-affirming features may actually reinforce delusional beliefs. Acting in accordance with your delusions – for example believing you are Jesus and blessing people – may provide distinct memories that confirm your beliefs further. Excessive thinking or ruminating about your special powers may actually be pleasurable. But the nature of that belief and its significance, such as saving humanity, may also drive excessive thinking about it, if failure to achieve it is stressful.[34]

What is it that causes people to develop delusions in the first place?

In normal circumstances, we assess our beliefs according to the world as we experience it. If something that we see or hear runs counter to what we believe, or what we expect to happen, then we question our beliefs or adjust them. For example, if I get the notion in my head that I am God, omniscient and omnipotent, my real-world experience of being unable to remember my password, or find my keys, quickly dissuades me of the truth of that idea.

There is a view of how the brain works that has become dominant in the world of cognitive neuroscience. It is that our nervous system is simply incapable of reconstructing the world around us on a millisecond-by-millisecond basis, dependent on the millions of data points that constantly flow through our eyes, our ears, our skin and other sensory organs. There is just too much information, too much computation, to cope with. Instead, the brain works as a prediction machine, and for the most part we see the world as we expect it to appear. It is only when there is discordance between what the brain predicts and what our senses tell us, that we need to adjust our expectations.[35] Our brains are largely configured to detect mismatches between our expectation of the world and actual events, seeking to learn from these moments. This process is not just limited to how we understand our external world, however. It is also how we comprehend our internal world: ourselves, our beliefs. It is how we learn, how we refine our understanding of the world we inhabit. If we believe something to be true, but our experiences inform us otherwise, we mould or adjust our beliefs.

For most of us, we are constantly amending what we predict of the world. The coat hanging on the door in the darkness of the night, briefly convincing us of an intruder in the room, before reason intervenes. Our political or social views, shifting throughout our lives in the face of new information and experience. In delusions, however, this does not happen. By definition, a delusional belief system holds firm in the face of overwhelming evidence that disproves it. For a delusion to develop, something about the delusional brain fails to properly assess the veracity of external information and update its expectations of the world. Those inconsistencies – that even as all-knowing God, I cannot work my hospital's IT system – do little to make me change my mind. (Or that Chris, despite being able to see his reflection, was not the most handsome man on the train.) The 'belief evaluation system' may be faulty in some way.[36]

This process of detecting when our expectations of the world are violated by external information is the ghost in the machine. It is at the heart of who we are, what we believe and how we perceive the world and our place within it. Intuitively, to me this process feels intangible: the point at which brain and mind most obviously diverge, despite my view that everything is *of* the brain. Yet there are indications that this 'belief evaluation system' does indeed reside within our brains, and in one particular region. If you scan healthy individuals and present them with scenarios where there is discordance between expectation and reality, the right pre-frontal cortex lights up, showing increased activity. When the same scans are performed on patients with delusional beliefs, this area produces much less signal.[37]

These sorts of scans suggest that this brain area is the seat of our belief evaluation system, although they do not provide definitive proof. But evidence also comes from other types of experiments, as we have seen throughout this book – experiments of nature. Occasionally, we see delusions arise as a result of brain injury or other damage, clearly visible on brain scans. Among these patients, with stroke or brain trauma, there is a common location of harm. These individuals almost always have injury to the right pre-frontal cortex, or regions intimately connected to it.[38]

Why delusions of grandeur?

This concept of a defect in reinterpreting our beliefs in the face of real-world evidence to the contrary may thus explain delusions, and hallucinations for that matter too. But why should some people experience delusions of grandiosity specifically, as opposed to delusions of other types, such as thought control or persecution?

The short answer is that we don't really know. However, there are some additional factors that may influence the flavour of delusions, and hint at the processes that influence the nature of delusions.

Grandiose delusions are strongly associated with lower anxiety and depression, and a positive evaluation of oneself; whereas persecutory delusions are linked to negative emotions and self-esteem.[39] The delusional brain therefore seems strongly influenced by emotional state. While the presence of delusions is determined by whatever pathological process induces this failure to adjust expectations, the nature of the delusional beliefs is mediated by underlying emotions. This would certainly explain why grandiose delusions are more often seen in the context of the mania of bipolar disorder, characterised by euphoria, excitement and increased energy, than other psychotic illnesses.

Brain studies also support the view that other factors influence the nature of delusions someone is likely to experience. In brain scans of people with schizophrenia, there are specific areas of the frontal lobes that light up much more intensely in those with grandiose delusions, compared to those with persecutory delusions.[*40] The normal functions of these brain regions, more active in grandiosity, may have direct relevance. In healthy individuals, these areas of the frontal lobes are suppressed when asked to judge others, but not when asked to judge or evaluate themselves.[41] In individuals without delusions, high activity within these brain locations therefore correlates with a focus on oneself, one's inner experience, at the expense of others. Low activity is associated with an attention directed towards others, and empathy for them; empathy by necessity requires a mindset that is outward-facing.

These differences in brain state – whether someone is prone to seeing the world from their own perspective, or from the perspective of others – may well explain why some patients with delusions have persecutory beliefs, while others have grandiose

---

* The medial pre-frontal cortex and the closely associated anterior cingulate.

patterns of thinking. If your brain is predisposed to delusional thinking, and you are attuned to focussing on others, you are likely to attribute your delusions to other people; hence themes of persecution, of being controlled, of having others transmitting thoughts directly into your brain. If your focus is directed upon yourself, then delusions too are likely to be directed towards yourself. These manifest as delusions of grandiosity or other delusions of self-significance, such as the false belief that irrelevant occurrences or details in the world relate very directly to oneself – for example, hearing comments on TV and believing they are directed entirely at you.[42]

So, delusional thinking results from a failure to modulate our thinking according to the realities of the world, due to impaired ability to detect discordance between reality and expectation. When our brains are more attuned to ourselves rather than others, the nature of those delusions are ascribed to ourselves, rather than to those around us – delusions of grandeur rather than persecution.

* * *

Having received the call during the journey from hospital informing Wendy of Chris's diagnosis of brain cancer, the couple were briefly left in limbo. Chris had been discharged on a Friday, so by the time they got home there was no one to contact. 'I had the whole weekend of thinking: "What is going on? Where do I go?"' recalls Wendy painfully. They found themselves a few days later sitting in an oncologist's clinic room. 'We literally said: "We don't know what we are doing here. We have heard nothing about cancer. Can I have another scan?" I was quite forceful, which irritated him straight away,' Chris tells me. The oncologist pointed to the scans, indicating the spherical abnormality. Wendy remembers the doctor pulling out a piece of paper, drawing Chris's brain, and describing the 'lesion' as an octopus, the head having been removed by the surgeons, the tentacles needing to be treated with radiotherapy and chemotherapy. Despite their protestations, and requests for another scan, within a few more days Chris found himself having radiotherapy and oral chemotherapy. 'I couldn't even see the fucking paper because my eyesight was gone at the time. I signed it [the consent form] and then we started the treatment,' he simmers.*

For the initial six weeks of treatment, Wendy and Chris recall him getting better and better, while those other patients around him deteriorated. As the steroid dose was weaned down, Wendy noted a steady improvement in Chris's mental state. Chris puts it more succinctly: 'It was like Darwin's evolution. I started as a monkey, and gradually became an upright human.'

Throughout this time, though, Chris's heightened self-belief continued. I ask them both if he thought he had special powers. 'It was very much like that,' Wendy tells me. 'It didn't matter what anybody said, or anybody did, he was better than them.' Chris confirms that he believed he had special powers insofar as, 'I understood everything. And they [other people] didn't. In terms of hypothesising about dark matter, dark energy, and what is entropy. It meshed between what my interests are and my accelerated view of things.'

He remained argumentative, challenging his doctors, railing against the cancer diagnosis. He asked for a second opinion. His behaviour rang alarm bells, and Wendy was told that Chris remained in denial about his imminent death. Chris's refusal to continue with further treatment, either radiotherapy or drugs,

---

* Chris later tells me that, in the course of his own investigations, he is unable to find a copy of this consent form in any of his medical records. When reviewing this manuscript, he writes that he was so intoxicated by the steroids, so impaired by the two operations, that he never gave his informed consent to adjuvant treatment. He would have preferred to take a 'wait-and-see' approach.[43]

led to attempts to section him under the Mental Health Act. He was referred for hospice care, and he tells me they again tried to section him, to initiate end-of-life palliative care.

It was on stopping his steroids, initially given for brain swelling, that Chris's behaviour dramatically improved, thus demonstrating the obvious impact of the steroids on his mental health. Wendy says: 'As soon as they stopped, his brain seemed to start opening up, and working better.' His complete self-regard lessened, and Wendy began to find him easier to deal with.

Grandiosity and the abnormal brain

As Chris illustrates, it is not just psychiatric disorders that give rise to grandiosity or frank delusions of grandeur, however. In his case, it is arguable if he was delusional – his heightened feelings of superiority are in themselves not delusions, although his conviction that he was the most handsome man on the train could be perceived as such, as does his belief that he could understand the world in a way that no one else was able to.

We have already seen that damage to the brain, through stroke for example, can result in delusions too. This is not at all surprising. Increasingly, our understanding of psychiatric disorders is that these are simply disorders of the brain characterised by chemical or organisational changes rather than wholesale visible destruction or degeneration of brain tissue. If changes to brain activity can cause these kinds of symptoms, then it makes perfect sense that physical damage to the brain could also result in similar alterations to our internal or external experiences.

A broad array of neurological disorders – tumours, Huntington's disease, inflammation due to viruses or auto-immune conditions, and traumatic brain injury – have all been identified as rarely giving rise to mania, in which grandiose delusions can occur. One report describes a fifty-nine-year-old man with Huntington's disease who believed he had special powers, and could control armies and the FBI.[44] Another paper, early in the AIDS epidemic, before even the causative agent of HIV had been identified, highlighted patients with AIDS who had grandiose delusions: one man believed that he had discovered a treatment for his condition which had been accepted for publication, and that he had cured himself of it.[45] The authors of the paper speculated that the psychiatric symptoms implied that the unknown AIDS pathogen may lead to brain involvement, something we are all too familiar with now.

Transient delusions of grandeur have also been reported after seizures. One twenty-five-year-old labourer was reported as suddenly behaving 'oddly', before believing that God was speaking directly to him and appearing in front of him, giving the patient omnipotence. These symptoms only lasted a couple of days. Initially he was diagnosed with a transient psychotic disorder, but further episodes and exploration of his symptoms led to the a diagnosis of epilepsy affecting a small region of his brain rather than its entirety, and anti-epileptic medication prevented further events.[46] This story is somewhat reminiscent of Sarah and her episodic morbid jealousy (see Chapter 4).

Indeed, causes outside the brain – conditions affecting other organs in the body or even external substances – that ultimately influence brain function, such as hormonal changes, vitamin deficiencies and drugs, can also do this. I can think of several patients over the years who, like Chris, have been given high-dose intravenous or oral steroids to treat an underlying neurological disease, becoming euphoric, sleepless, full of energy and completely grandiose, only for their psychiatric symptoms to dissipate after the drug has been discontinued. That steroids can induce neuropsychiatric side effects has been recognised for decades. These drugs are associated with a wide range of symptoms, from insomnia and anxiety, to depression, hypomania (a less severe form of mania, characterised by an elevated mood, increased energy, inflated self-esteem) and frank psychosis; these more serious manifestations are usually seen at very high doses. Such effects are not uncommon, and some studies suggest that up of 60 per cent of patients will exhibit some sort of psychiatric disturbance with steroid treatment.[47] Little is known as to how precisely these drugs induce neuropsychiatric side effects, although steroids are known to have a broad range of effects on neurotransmitters, as well as influencing concentrations of receptors. Rarely, even in the absence of steroids as a treatment, tumours of the adrenal gland, leading to the production of excessive natural steroids, can cause this.[48] These 'organic'* causes of psychiatric disease account for a significant proportion of patients presenting with mania – just under 5 per cent in one study.[50]

These sorts of cases, where a drug or structural changes gives rise to delusions, show us how fragile our neurological systems are when it comes to our grasp on reality. They also show us that, as with emotional states and mindsets related to ourselves or others influencing the nature of our delusional beliefs, certain drugs like steroids seem to have predilection for inducing chemical changes within our brains that prime us for grandiosity, for self-importance. In Chris's case, his superhuman insights into the nature of the universe and his dazzling good looks, for example, are down to the high doses of steroids.

---

* The term 'organic' is still used to describe medical or neurological causes of psychiatric symptoms, as opposed to mental or psychogenic causes, despite it being removed from classifications of psychiatric disorders in the mid-1990s.[49] Of course, this distinction is becoming less and less applicable as we begin to understand that considering psychiatric disorders as diseases of the mind rather than of the body is incorrect, but the term remains in widespread use, even by psychiatrists.

Chris feels that his grandiosity was entirely triggered by the medication that he was taking, and that he normalised as soon as he stopped them. Wendy, however, has a different viewpoint: that they do not explain everything. While she agrees that he got much better, she is very clear that even now, eight years later, he has been left altered. He continues to exhibit a heightened sense of self, of being right when everyone around him, including his wife, is wrong.* 'When he was on the steroids, he got this overriding urge that he was the be-all and end-all. He is definitely left changed, without a shadow of a doubt. His ways of thinking are very different now.' A different Chris to the man she married and lived with for decades before his illness.

While they both play down his behaviour, it has obviously taken its toll. 'It was very difficult because it came to the point where I saw myself not as a wife anymore. He only saw me as a carer, nothing more than a carer. And if the caring wasn't up to what he was needing at that time, then . . . yeah . . . ' she peters out. Chris believes that he did indeed begin to see her as his wife again, but was trying to re-establish his self-worth in their relationship. At one point, however, it became so unbearable for Wendy that she left the family home with her daughter, with the intention of leaving Chris for good.

Eight years later, now in his mid-fifties, Chris remains physically well, his most recent brain scan showing no evidence of a tumour. For someone with a very aggressive large brain tumour, who never completed treatment, this is highly unusual. Chris remains utterly convinced that he never had cancer, that it was an abscess all along, as they were originally told by the

---

* On review, Chris later adds: 'Given his numerous audited "life-critical" engineering supervisory roles, this trait, along with logical analysis, could be construed as a positive.'

neurosurgeons at the point of discharge. Indeed, to my eye, his scans look like the classical appearances of an abscess, and I can see why this was the initial diagnosis.

Wendy and Chris think they know where the confusion may have arisen. In the same bay of the neurosurgical unit, Chris met another patient named Chris, whose surname differed from his by only one letter, and whose date of birth was very similar. This man knew he had a recurrence of a GBM. 'We joked that we were brothers from another mother,' Chris says, both men struck by the similarity in their names and birthdays.

However, documentation Chris later shows me confirms that the biopsy from his second operation was reviewed at two separate hospitals, in two independent multidisciplinary meetings. The conclusion from both teams was that this abnormality was indeed an aggressive cancer – GBM. In the context of his extraordinary survival, some eight years after being told of his impending demise, his biopsy is being reviewed again. I can certainly understand why he is fighting so hard to get answers though, to interrogate the healthcare system that he and Wendy feel let him down so badly.

Chris tells me his recent scans show signs of radiation damage, and in the last year or two he has developed seizures. Although both he and Wendy smile and laugh as they tell me their story, they are obviously traumatised, and their anger boils barely concealed. In Chris's many emails to me, he seethes with indignation that no one listened to him, that his concerns were ignored. That he appears to have been right all along. That his doctors were 'idiots', 'imbeciles', who discounted his views due to his lack of formal education. 'I have found that the most educated university-trained people are the most ignorant people. Blinkered, rigid,' he says. He seeks closure, an apology, a recognition of his views.

When we discuss the behavioural aspects of Chris's case, I put it to him that in my mind he has always had subtle personality traits, even before his illness, that in some respects represent a form of grandiosity, a supreme self-confidence in his abilities, although Chris might argue that his beliefs are based in reality. The steroids, however, induced an altered brain state, bordering on delusional, but certainly a heightening of these views. I suspect that his unusual behaviour on the steroids may have contributed to the fraught relationship with his oncologists when he was questioning his diagnosis.

Years after his illness, his personality remains changed from the brain damage sustained due to the underlying disease, the surgery and the radiation accentuating his longstanding tendencies. When we talk, Chris seems to agree with me when we discuss this in person. However, in the subsequent revision of my conclusions, as we go through what I have written, it is clear that Chris is less convinced of this. While he accepts that he has been fundamentally changed by his medical condition and its treatment, he seems to put more emphasis on the steroids and his experiences while on them. And I get the impression that there remains some discordance between his views and those of his wife about the degree of his recovery. Among the emails to me,* he later writes:

When we spoke, I advised you that you should remain cautious about attempting to establish a causal link between brain damage and episodes of . . . 'grandiosity'. Although we could have a separate debate about the nature of personality, and how you hypothesise that these changes come about, or show themselves, as seemingly entirely

---

* At certain points, Chris sends me multiple emails a day, with subjects ranging from 'Accelerated Cognitive Repair Through Rationalised Stress Management', to 'Medicinally Induced Cognitive Entrapment', and 'What are Dreams, and Why Do We Dream?' They seem to hover between grandiosity and the outpourings of a highly inquisitive mind.

new traits, after an impactful cognitive event, in my humble opinion it is not a linear 'cause and effect' relationship. From my own – and dare I say invaluable – silent-witness account, the role of prescribed dexamethasone [the potent steroid prescribed], cannot be overlooked. What this steroid does, is magnify existing positive and useful aspects in some personalities to such an extreme, that they are reasoned to be entirely new to the casual observer. When, in fact, they are not . . . Steroids merely induce, temporarily these Jekyll and Hyde changes to existing characteristics. Which is why it is so important that diagnosing clinicians, need to be careful not to presuppose or mislabel patients unfairly . . . Especially those clinicians who, as a result of their lack of psychological understanding, deem a patient to be 'in denial' about their own imminent death.

For me, Chris's case illustrates the blurred boundaries between personality and psychiatric disease, where the worlds of psychiatry and neurology collide, the meeting of mind and brain. As Chris says: 'What we are, who we are, is no more than a mass of electrical signals in the brain.' The outward manifestation of the confluence of our brain structure, function, chemistry and genes.

It also demonstrates the fine line between the benefits and detriments of pride, grandiosity or self-belief – however you describe it. Chris's own faith in his intellect, his self-certainty, has driven his career and his academic interests. It is the basis for his relentless determination to clarify what has happened to him, to rationalise his perceived betrayal by the medical system, and to seek justice. There are clearly upsides to this character trait, but there are harms too. The price was almost his relationship with his wife. And I cannot help but feel that his dogged resolve to pursue closure, to address the uncertainties surrounding his diagnosis and treatment, comes at great cost.

8.

Free Will

The priest knows, as everyone knows, that there is no longer any 'God', or any 'sinner', or any 'Saviour' — that 'free will' and the 'moral order of the world' are lies: serious reflection, the profound self-conquest of the spirit, allow no man to pretend he does not know it.
                                 Friedrich Nietzsche, *The Antichrist*

In 1978, Robert Alton Harris, then twenty-five years old, asked his younger brother Daniel, aged eighteen, for help in carrying out a bank robbery.[1] Daniel had stolen two guns from a neighbour's house in Visalia, California, and the next step was to steal a car for the bank job. The brothers came across two teenage boys, John Mayeski and Michael Baker, parked in a green Ford LTD outside a fast-food restaurant, eating their lunch of cheeseburgers, and saw an opportunity. Mayeski and Baker had planned to spend the day fishing to celebrate Mayeski's recent acquisition of a driver's licence. Instead, the Harrises forced the two sixteen-year-olds to drive them to a remote location, Miramar Lake, at the barrel of a 9 mm Luger, Robert commandeering the Ford and Daniel following behind in another car.

What happened next is not entirely certain. According to the prosecuting attorney record, the friends were forced to kneel, and began to pray. Robert apparently told Mayeski and Baker:

'Quit crying, and die like men!' before shooting them both multiple times. Newspaper articles of the time report the story a little differently. Robert convinced the boys that they would be unharmed, that the Harris brothers would return the car to them and give them a share of the loot. As Mayeski and Baker began to walk away, Robert pulled out the gun and shot them from behind, to his brother's total surprise.

In one report, from the *LA Times*, relying heavily on Daniel Harris's description of events, Robert slowly raised the Luger as Mayeski began to walk away, shooting him in the back. Baker made a run for it down a hill, but was soon chased down, and Robert shot him several times. When Robert Harris made his way back up from the valley, he found Mayeski lying on the ground but still alive. According to Daniel, Robert knelt down on the ground, held the Luger to Mayeski's head, and fired. Daniel reported: 'It was like slow motion. I saw the gun, and then his head explode like a balloon . . . I just started running and running . . . but I heard Robert and turned around. He was winging the rifle and pistol in the air and laughing. God, that laugh made blood and bone freeze in me.'

After the slaughter, the brothers drove off with the guns and the remainder of the dead boys' lunch. About fifteen minutes after the murder, Robert took the cheeseburgers out and polished them off. Daniel, meanwhile, reported feeling sick and ran to the bathroom. 'Robert laughed at me. He said I was weak; he called me a sissy and said I didn't have the stomach for it.' Robert allegedly was elated, and joked that it would be funny to dress up as police officers and inform Mayeski's and Baker's parents that their sons had been killed.

Shortly after the double murder, the Harris brothers robbed the bank opposite the car park where they had abducted the two boys, fleeing with about $2,000. But they were arrested within the hour; a witness had followed them home and called the police. In a twist of fate, it was Michael Baker's father, a police officer, who apprehended the brothers, not yet aware that his own son had been killed.

In March 1979, the younger brother, Daniel, was convicted of kidnapping, and sentenced to six years in state prison, eventually released in 1983. The elder brother, Robert, was convicted of two counts of first-degree murder and two counts of kidnapping and was sentenced to death. He met his fate in the gas chamber at San Quentin prison, on 21 April 1992 (his last meal a 21-piece bucket of KFC, two Domino's pizzas, a bag of jellybeans, washed down with a six-pack of Pepsi and a pack of Camel cigarettes). Apparently, the inmates at San Quentin had planned to celebrate his execution, so awful was he perceived to be by the most hardened criminals in the one of the harshest penal systems in the world.

When you hear the story of Robert Alton Harris, it is hard to imagine how anyone can be such a 'monster' – so devoid of morality, evil beyond compare. To execute two sixteen-year-old boys in cold blood, to do so with glee, with jubilation, without a flicker of remorse. He could have easily made different choices, both on that day and in earlier times, and he himself led his life to its ultimate destination. An evil man meeting a fitting end.

But there is another perspective on the nature of this man. Robert was born in Fort Bragg, North Carolina, the fifth of nine children. His parents were both alcoholics, and he was born two months prematurely, when his father kicked his mother in the stomach, convinced that the unborn child was the product of infidelity. Robert had a speech impediment and learning difficulties, attributed to the toxic effects of alcohol on his developing brain. His father beat all the children mercilessly, and sexually abused Robert's sisters. His mother would deny Robert any physical contact, and on one occasion apparently bloodied his nose when he tried to touch her. Aged fourteen, he was in a detention centre for trying to steal cars, and was raped several times, subsequently attempting suicide by slashing his wrists. Shortly after being released, aged nineteen, he was killing and torturing dogs, cats and pigs, with pellet guns, darts, knives and sticks.

In this context, it is easier to be a little more ambivalent about Harris. Yes, a monster. Yes, abhorrent. But was he the architect of his own life – or was it determined by external factors? Was it inevitable that the combination of genes, the stunting effects of alcohol on his brain while in his mother's womb, his upbringing, his experiences in detention, determined that he would become the cold-blooded murderer he was?

While Harris's case is obviously extreme, it illustrates the arguments about what makes us the people we are, the nature and nurture that act like two hands moulding our brains from clay. And crucially, to what degree our actions result from free will, the choice to proceed down one path or another.[2]

For if you believe that the brain is the origin of our personalities and our character traits, the basis of our decisions, be they good or bad, then it is arguable that much of what defines us is outside of our control. We do not choose our parents, and therefore have no influence on our genes or early upbringing. Equally, the vicissitudes of life tend to drag us in the current, rather than us swimming strongly in the stream.

But if we are all passive bystanders to our own actions, what then happens to responsibility? Can we ultimately be responsible for something we have limited or even no control of? Without the freedom to pick a particular path, to choose whether to indulge in wrath, gluttony, lust, envy, sloth, greed or pride, can these thoughts or behaviours really be considered illustrative of our moral value as individuals?

What then of free will?

The death of free will

The terminology of modern medical English is peppered with words originating from other languages. Predominantly, these are Greek or Latin, words like diarrhoea, catarrh, diabetes or primigravida. In fact, approximately 95 per cent of English medical terms are borrowed from or constructed from Latin or Latinised Greek. There is the odd word derived from Arabic – alcohol – or Italian, like malaria, influenza or quarantine. French also makes several appearances, deriving from the Middle Ages, as in malaise, leprosy or polyp, or from the Age of Enlightenment, with words such as oxygen and déjà vu.[3]

But there is another language that has also contributed. It conjures up in my imagination Freud and his compatriots at his salon in Vienna, at their Wednesday Psychological Society meetings, sagely stroking their beards as they discussed their latest theories on the human condition. German words like *Angst*, *Merkwelt* (way of looking at the world) or *Weltschmerz* ('world grief' or weariness). Psychological terms describing complex states of the human mind and soul. Or words such as *Zeitgeber*, an external cue for the circadian clock. Perhaps due to my innate prejudice borne of my German roots, in my mind it is the language of precision, exactitude, maybe even pedantry. *Bereitschaftspotential* is such a word, describing a precise observation of the brain, but with uncertain implications.

The origins of the *Bereitschaftspotential* are rather picturesque – in the garden of the 'Gasthaus zum Schwanen' at the bottom of the Schlossberg hill, to the east of the old town in Freiburg. The German city, at the southern end of the Black Forest, is famed for its university. In that garden, on a sunny spring day in 1964, Hans Kornhuber, a neurologist, and his doctoral student, Lüder Deecke, had gone for lunch. As they sat and ate, alone in the sunshine, they talked of their frustration at the state of brain research, and the apparent focus on the study of the passive brain, how it detects stimuli and responds to them. They agreed that a more exciting avenue of research would be to attempt to understand brain mechanisms of self-initiated movements, of how the brain generates actions – a refocus away from how we perceive the world towards how we have agency, to act on our own behalf. And so the pair planned and executed a study in their laboratory, using twelve healthy volunteers, students at the university. Little did they know how explosive their findings would ultimately be, leading people to fundamentally question the nature of what it is to be sentient, to be human.

Kornhuber and Deecke got their volunteers to perform hundreds of movements while attached to an EEG machine, recording their brainwaves. Sometimes these movements were active, initiated by the volunteers: pressing a button, closing a rubber ball in their hand or pulling on a string by flexing their wrist. At other times, the movements were passive, and tried as much as possible to simulate the same movements – for example, by the examiner closing the volunteer's hand around the ball, or pulling on a rope to flex the wrist. When Kornhuber and Deecke combined the brainwave traces of hundreds of these movements, and averaged them out, they found something curious. Time-locking these EEG traces to the movements, they found an electrical signal, most evident over the opposite hemisphere from the movement, but spreading to both sides of the brain, preceding active but not passive movements. These changes in electrical activity were detectable on average 1–1.5 seconds before muscle activity started, and the researchers proposed that this signal represented an early preparatory process within the brain prior to active movement. This signal was termed the '*Bereitschaftspotential*', or readiness potential, and was proposed as essentially signifying the brain preparing to make a movement of its own volition: a marker of planning to move.

The *Bereitschaftspotential* (BP) has entered into usage in the clinical neurology world to some degree, and is occasionally helpful in discriminating jerks or twitches that have a voluntary or involuntary origin.[4] But its relevance to all of us, what it tells us (or may tell us) about our own nature, is somewhat more dramatic. Its discovery triggered a neuroscientific earthquake.

Following its description in 1964, researchers began to wonder why there was such a delay between the apparent planning of an action, marked by the BP, and its execution. In 1983, Benjamin Libet, a neuropsychologist working at the University of California, San Francisco, asked five volunteers to sit in front of a screen while monitoring their brainwaves with EEG electrodes.[5] On the screen was displayed a clock face, with a dot moving round it at speed, roughly one full rotation of the clock every 2.5 seconds. These volunteers were asked to make a hand movement at a time of their choosing, without planning, but simply to note where the dot had been when they first experienced the urge or made the decision to move. Each subject did this hundreds of times, and as with Kornhuber and Deecke's study, the EEG was averaged and timelocked to the movements, to assess the BP. To Libet's surprise, the urge to move, or the decision to move, occurred only 0.2–0.3 seconds before the movement itself, quite some time after the BP had arisen. Libet and his colleagues summarised this as: 'Cerebral activity initiates the voluntary act before reportable conscious intention appears.'[6]

It is worth reading this statement again. This is as close to a bombshell as one could imagine in the dry world of the scientific literature. 'Cerebral activity initiates the voluntary act before reportable conscious intention appears.' Essentially, what Libet was saying is that processes in the brain are making decisions long before we are aware of making them. And to take that forward, as many researchers who have replicated his study since have done, that there is no such thing as free will. That our perception that we have freedom to make decisions is nothing other than an illusion; that our perception of having a conscious choice to act is a grand deceit. We may think that we are choosing to turn right or left, to order pizza or pasta, to say yes or no, but actually processes in the brain have made those decisions for us long before we are aware of having made them.

Libet's results were earth-shattering. His study has been cited by other researchers almost four thousand times. Some have likened this apparent illusory nature of free will to a car turning, and the indicators switching on prior to the turn. An observer may think that the indicators dictate which way the car is turning, but in reality, it is the person in the driver's seat, who has made the decision before indicating. The indicators are an 'epiphenomenon'* of the true nature of decision-making. Our unconscious brain is, in fact, the driver of that car. The indicators are purely when we become aware of that decision. A shocking conclusion, shaking the foundations of our understanding of ourselves, and our perception of our internal worlds.

In fairness, Libet was rather more conservative in his view, and did not extrapolate his results to the degree of denying the existence of free will. He argued that even if the BP represented making a decision, we could still exert free will through veto power, choosing not to act on that decision: 'free won't' rather than 'free will'. But his paper was the lighting of the touchpaper, the key finding in support of the deterministic view – that

---

* An epiphenomenon is defined as something that can be detected simultaneous with another phenomenon but is not causative. In medicine, when two phenomena – such as symptoms, diseases, or test results – occur together, it might be because one is causative of the other, or they may simply be associated. A good example is the fever associated with infection. If someone is very unwell with pneumonia and a fever, you could assume that it is the fever making them unwell. In truth, however, it is the bacteria in their lungs causing them to be sick, and also causing the fever.

everything is predetermined, whether it be by fate, or physical laws or rules – in this setting, decisions are predetermined by the nuts and bolts, the circuitry, anatomy and neurochemistry, of our brains.

The fightback

Despite the huge impact that Libet's results had on neuroscientists and philosophers alike, the fightback against both what the BP means and the nature of free will started almost immediately.[7] The character of the BP, and the tiny size of its electrical signal in relation to the background rhythms of the wakeful brain, means that it can only be detected by averaging out many recordings, and some researchers very early on pointed out that this process may well give rise to signal of no clear meaning – a result of experimental technique rather than reality. Others argued that the urge or intention to act was not an all-or-nothing event of sudden onset that could be marked by a dot on a clock face; rather it built up slowly and the subjects were only marking the peak of that 'urge'. Some neuroscientists took issue with the view that Libet's experiment was a good replication for voluntary movement in general. The simple act of bending your fingers or wrist was not in any way representative of actions of wider meaning, of decision-making in the real world, and that Libet's volunteers were simply obeying instructions: 'These simple movements are made voluntary, but the will acts here only as a trigger. Willed intention is more important in goal-directed and complex movements such as writing.'[8]

But ultimately, the fundamental question is not about limitations of Libet's methodology to accurately record the BP. It is the assumption that the BP is the *cause* of the action in the first place, that it represents the decision to act. Because, if it is not – if the BP is as a result of something else – then the implications of Libet's experiment on concepts of free will simply evaporate.

Research in the last decade or so has interrogated this assumption. A number of strands of evidence have cast doubt on the view that the BP represents brain activity preparing for movement.[9] Experiments have shown that occasionally the BP can be seen before unconscious or involuntary movements, or even before decisions that do not involve movement at all. And conversely, that sometimes the BP is not detectable at all before deliberate choice movements.

So, if the BP is *not* the brain preparing to make a movement, then what does this averaged-out brain signal actually represent? A number of different research groups have proposed a rather more prosaic explanation.

In real life, we base our decisions, our actions, by weighing up internal and external information. We see clouds in the sky and decide whether or not to go back for an umbrella. We look at the rows of fruit and vegetables at the supermarket, and assess the ripeness, the quality, the price and our personal likes and dislikes, before choosing which to buy. But for Libet's subjects, there was none of this. Just a simple decision as to when to act, whenever it occurred to them to move their hands. A question of when to do something, not what to do. So perhaps the BP does not represent the first unconscious preparation of choice, but something else entirely.

Instead, the BP may simply be a product of averaging the electrical activity of millions of individual nerve cells, an approximation of natural fluctuations of brain activity. If that is the case, then maybe those spontaneous acts – of pressing a button in response to Libet's clock, or moving fingers in Kornhuber and Deecke's original experiments, devoid of external triggers and preordained – is prone to occur when the brain drifts into a state more likely to result in the initiation of movement.[10] Essentially, when faced with a meaningless task, one without a clear goal and without clear inputs on which to base a decision, the choice to act may simply be dependent on the brain state reaching some sort of tipping point. To act in this context may purely result from a breaking of the stalemate of decisional equipoise. And the BP is the electrical signature marking the end of ambivalence.

Instead of being the death knell of free will, the BP may therefore simply be a marker of a brain state that promotes us to move in the absence of any other reason to do so: a highly specific situation rather than something that can be extrapolated across the breadth of the human experience. There is something rather poetic in all this – that our views on the illusory nature of free will may actually be driven by the illusion of the BP being a decision, a choice to act: as one neuroscientist wrote, 'It is ironic that the "obvious" conclusion drawn from Libet's experiment that conscious intentions are an illusion will have to be replaced with the realisations that the readiness potential itself is an artifactual illusion.'[11]

The death of free will may have been announced prematurely, then. There are other experiments, however, studying brain activity more directly through electrodes implanted deep within the brain,[12] or using cutting-edge scanning techniques.[13] These too show changes in the brain before any conscious decision is made, in one case up to ten seconds before. Criticisms have also been levelled at these studies. That the tasks are not representative of free will, that decision-making is an inherently gradual process; that these signals are a marker of a precursor process that informs the conscious decision to move.

The neuroscientific arguments about the existence of free will rage on, as do debates on whether Robert Alton Harris was truly a 'monster', free to choose the path of murder, or whether he was a victim of an abused and damaged brain. I suspect these questions will be with us for some time, without definitive answers.

The reluctance to give up on free will also has other origins of course. You are free to read this sentence, or close this book and get a cup of coffee or a glass of wine. You can choose whether to cook dinner tonight, or to order takeaway. To punch someone in the face during an argument, or to walk away. What these studies imply is that this freedom does not exist, that processes within our brains determine our choices before we have any conscious awareness of them. And that concept – that what we perceive to be our freedom to choose, freedom to act, freedom to think, is simply a mirage – runs counter to every aspect of our conscious experience. It is an appalling thought that we are simply machines, predestined (however you define destiny) to act and choose in a particular way.

The lines between normal and abnormal – sad, mad or bad

Let us set aside the concept of free will for the moment, whether we have it or whether it is simply an illusion, and the implications of its absence. The individuals described in the pages of this book illustrate that our behaviours can be influenced by changes in our brain structure, brain chemistry and brain activity. If a change in behaviour can be as a result of a change in the brain, then it clearly indicates these thoughts and actions are of the brain too.

Many of the people I meet in clinical practice exhibit patterns of behaviour that are deemed as having 'negative moral value' – actions that are bad or 'evil', over which they have no control.

The person who becomes frankly psychotic and violent after a seizure; the individual who becomes hypersexual when given a medication. For these people, it is obvious that external factors, beyond their control, have precipitated deeds that are not intrinsic to them as human beings, that do not define them. But what about others, where such a direct causality of extrinsic factors resulting in malevolent or harmful acts is not so apparent?

History tells us that much of what we have previously determined as being derived from our 'soul', has a basis in cold, hard scientific fact. In the past, the physical was defined as something we could see, with the naked eye, on the living body or inside the human corpse. Then, with the advent of the microscope, many diseases previously undetermined acquired a physical basis, and diseases such as malaria could be visualised rather than be considered 'bad air'. But there has remained this view that, on the one hand, there is the neurological – what can be clearly defined as damage to the structure of the nervous system, be that macro- or microscopic, or somehow else objectively determined, like the abnormal electrical activity in the brain that constitutes epilepsy. On the other, there is the world of the psychiatric, the psychological – a more ephemeral, spiritual realm – the sphere of the behavioural, the emotional. There is a division between brain and mind.

This division is not static, however. It relies on our ability to detect and define physical change, as with the development of the microscope, or our advancement in imaging technology. Consider many of the patients I have described, with profound changes in behaviour as a result of tumours, inflammation, infection or strokes. In the days before X-rays, almost all of these people would have been diagnosed as 'sad, mad or bad', as having disorders of the mind or soul, unless they had the misfortune to die and have their brains studied at autopsy. With the development of X-rays, a few of these individuals with large tumours may have had a physical cause found. Then with the arrival of CT, another tranche of patients with large strokes or wholesale brain inflammation, would have shed the label of a disordered mind or spirit. And finally, with the advent of MRI and its improved resolution, almost all these individuals would have received a neurological diagnosis.

But this process of discovery, of finding these physical causes of alterations in personality, emotions and behaviour, does not halt here. What has happened in the world of neurology is now also happening in the worlds of psychiatry and psychology. As technology advances, as our ability to look at genes and the structure and function of the brain progresses, the worlds of the brain and mind are beginning to coalesce even further. The genetic contributors to serious mental health disorders such as schizophrenia and bipolar disorder are now being unravelled. The changes in brain circuitry and function in depression and other psychiatric diseases are beginning to be understood. The role of physical inflammation in depression, of childhood trauma in mental ill health, even the neurobiological basis of previously termed 'hysterical' or 'psychosomatic' disorders[14] – all of these are pointing towards the view that there is little or no distinction between the physical and the mental. When it comes to both the neurological and the psychological, they all originate in the brain. And by extrapolation, the emotional, the behavioural, they all originate in our brains too. They all have a basis in the physical. And if that is the case, it further blurs the boundaries between 'sad, mad and bad' – psychotic illness, other mental disorders, 'evil' – and indeed what we consider physical disorders of the brain.

There are of course dissenters to this view. Even among my own neurology colleagues, I have heard some say that their belief is that, even if we were to understand the brain at the level of every single sub-atomic particle, we would still not fully comprehend the mind. At the moment it is clear we are a long way from proving or disproving this viewpoint.

Some researchers urge a degree of caution when interpreting some of the studies I have shown you, in particular functional MRI. Functional imaging permeates the world of neuroscience research. By looking at patterns of blood flow within the brain, as a proxy for brain activity, the tendency is to ascribe those brain areas that show differences in particular tasks as being the anatomical origins of specific functions or traits. Critics of this research methodology point out that these scans are not a direct measure of neural activity, and that the resolution of this technique is limited to blood-flow changes. That the conclusions of these studies are only as good as the hypotheses behind the tasks being undertaken by the subjects in the scanner. And that these sorts of studies suffer from a 'replication crisis', where different research groups are unable to achieve the same results. Furthermore, in neuroscience there is a wholesale move away from considering blobs of tissue within the brain as being the 'love' centre, the 'face recognition' centre, the 'word production' centre, and an appreciation that brain functions are reliant on networks of blobs, connectivity between different brain regions. Our cognition, emotion, behaviour – all of these facets of who we are – are not dependent on individual small clumps of neurones, but instead are defined by an intricate dance between all these areas of the brain working in unison.

But this is the beauty of disease, at least from a scientific perspective. It does not rely on extrapolation of data, on proxy markers, on statistical significance. Instead, it relies on strokes, tumours, degeneration, inflammation or injury. If a specific brain region is damaged by disease, and the function or character of that individual is altered, then it clearly tells us that this part of the brain is fundamental to that function. It does not necessarily tell us that this is the 'grandiosity blob', or the 'envy centre', but it does inform us that this is an intrinsic part of the neurological machine that gives rise to these functions. Like a kitchen appliance in which a component burns out, and one particular programme stops working. That component is not necessarily entirely responsible for that programme but is crucial to it.

I stress, I am not saying that our affect, our character, our actions, are all predestined at birth; that we are unchanging and unchangeable. The brain is obviously a plastic structure, constantly changing and adapting. This is at the core of us learning facts, tasks, skills. It is the basis of recovery after stroke or other brain injuries. The brain is malleable in the furnace of our lives, forged by our experiences, our environment and our actions, as I have shown. But its foundations are our genetics, our anatomy, our chemistry.

If we accept that changes in brain structure, genetics, life experiences, physical and psychological trauma, can all rob some people of free will, is it that much of a leap to say that this is the case for all of us, to some extent? As with diagnostic shifts between what constitutes a physical or psychological disorder, medicine is full of examples of diagnostic creep – where a disease, and its underlying cause, is identified in a few extreme cases. With time, we look for this disease in milder forms, and find it, showing that that original cohort of patients is just the tip of the iceberg when it comes to that particular illness. As an example: cholesterol used to be seen as a problem for small groups of individuals with hereditary forms of disease, who died of strokes and heart disease at a very early age. Nowadays, of course, we understand that this is not just an issue for a few families, but that it has implications for us all.

Similarly, if we find genes that are associated with major depression, I suspect we will find that some of these genes will be associated with less severe depression, or even a generally gloomier outlook on life. The effects of these genes may be smaller if you are just a bit of a pessimist compared to someone with life-threatening depression, and those genes may need an interaction with your life experiences, or other factors to mani-fest as a depressive illness. As we have seen, this is already the case in anger, with some genetic variants predisposing people to violence, and these genetic factors interacting with our envi-ronment. Two individuals undergoing the same experiences may be influenced in very different ways, their brains liable to vari-able extents depending on their genetic, social, anatomical background. In neuroscience, as in other areas of medicine, disease or dysfunction is recognised to be a combination of biological, social and psychological factors. Ultimately though, there will be factors beyond all our control that define who and what we are.

In light of all this, then – our ever-advancing comprehension of the factors that influence our character, our choices, our behaviours – where do we draw the line between disease and good health, or whether someone is 'sad, mad or bad'? Who should be the arbiter of what constitutes the intrinsic nature of a person, or the effects of genes, anatomy or environment? Is it the judge, the jury, the doctor, the philosopher or the priest?

Without free will, what about sin?

If you take the view of Libet's proponents – that free will essen-tially does not exist, that our choices are not real choices, that we are simply at the behest of the machinations of our brains, and the factors that influence the machine – then there is an inevitable conclusion: there can be no moral responsibility. To be a moral agent requires you to reflect on your situation, to form an intention and then to act upon that intention. But without the ability to reflect and form an intention to act independent of the structure and function of your brain, without free will, you cannot have moral responsibility. Robert Alton Harris, the cold-blooded murderer of two innocent boys: a monster devoid of any ethics, or a product of his nature and nurture? Of course, the neuroscientific community is far from unified when it comes to whether there is a neurobiological basis to free will. Even Libet himself, decades after his famous experiment, wrote: 'In an issue so fundamentally important to our view of who we are, a claim for illusory nature [of our free will] should be based on fairly direct evidence.'[15]

Like neuroscientists, the philosophers are also far from consensus. On one end of the philosophical spectrum are the hard determinists, those who reject the notion of free will. That all decisions are not really decisions, but that only a single decision made by a moral agent is in reality possible, based upon their inherited character, life history and environmental stimuli. From a philosophical perspective, this is not a new concept. The ancient Greeks, like Socrates and Strato of Lampsacus, expressed views that what appear to be choices are actually obligate decisions, whether originating from the mind or some unconscious divine power. So too does the classical Hindu text, The Bhagavad Gita: 'They alone truly see who understand that all actions (of the body) are performed by material nature, while the self actually does nothing.' The philosopher Baruch Spinoza, writing in seventeenth-century Amsterdam, stated: 'Men are deceived because they think them-selves free . . . and the sole reason for thinking so is that they are conscious of their own actions, and ignorant of the causes by which those actions are determined.'[16]

Others take the opposing view, that free will and thus moral responsibility exist; that any individual has the ability to take more than one course of action under any specific set of circumstances. Some libertarians stray into the world of dualism: that processes in the brain that lead to actions have explanations other than physical, that there is a mind or soul separate from the brain. This is not a view I subscribe to. But other libertarians propose that the soul or mind is not fundamental to free will, but that there are physical processes that may have more than one outcome: for example, the probabilistic behaviour of sub-atomic particles in quantum mechanics theory.

In between, sit the compatibilists, who propose that free will and determinism are not mutually exclusive and can be reconciled. They argue that it is possible for individuals to possess free will, even if their actions are determined by internal processes and external circumstances. For compatibilists, people can be held morally responsible even in the face of determinism, where their actions are a result of their own desires and intentions. For them, the definition of free will is not dependent on those external or internal forces that define us. Rather, free will is the ability to act in accordance with one's own values and wants, without coercion or manipulation.

I am certainly no more likely to give a definitive answer as to free will than the philosophers or the neuroscientists dedicating their lives to this area of research. But what I can say is that I have met many people (a few of whom I have introduced you to in this book) in whom free will – the ability to choose, to make a decision, to perform an act – seems to have been attenuated or eliminated by disease or injury; whose behaviours appear to have been driven by changes in the brain rather than their own characters and values. There are some in whom the issue is genetic, determined at conception, and others in whom environmental factors have fuelled particular patterns of conduct.

The view that there is no such thing as free will is at the extreme of the argument, but at the very least, there are many factors that we *as individuals* cannot fully control. Either way, our ability to decide, to perform moral acts, may not be entirely of our own making. And if that is the case, it has important implications on our views of morality, of good and evil. For if our decisions, our choices to act or not act, to do something harmful to ourselves or others, are indeed wholly defined or partly influenced by the machinations of our brains rather than this ephemeral concept of our minds or souls, then to deem us as being inherently good or bad individuals is without meaning. Perhaps it is not a matter of being good or being evil, but just of being.

On a scientific level, I am very comfortable with this deterministic perspective. It strongly chimes with my neurological world view: that we are entirely a product of that organ comprising just over a kilo of jelly that resides in our skulls. That our opinions, our actions, our personalities – essentially all that makes us human – are simply a function of neurones, of neurotransmitters, of trillions of synapses. That there is no soul, no intangible mind divisible from the brain. I am acutely aware that my own opinions would be influenced by those factors influencing my own brain function.

On an individual level though, I squirm at this. My own personal experience convinces me of my ability to decide whether to turn left or right, to eat a chocolate bar or an apple, to say something nice or nasty. To consider free will as simply an illusion runs contrary to every fibre of my being. It is difficult to believe that I might be completely deceived by my own mind, and that this too is the case for everyone else.

* * *

I started this book with my own family's experiences of evil, of the horrors of the Holocaust, the wholesale genocide of Jews, Roma, Sinti, Slavs, homosexuals and many others, by a group of people who considered themselves morally upright, who justified their actions, at least to themselves, in some form of theology. To deem someone like Hitler as having no individual moral worth, no personal responsibility, makes me viscerally uncomfortable, and I am not sure I can square that thought in my own mind. I can rationalise the underlying emotional and behavioural contributors to my family history. But to explain or excuse on an emotional level, even to understand, is simply a step too far.

Nor do I find it easy to forgive and comprehend those people in the world today committing terrible acts of violence, slaughter, rape and hatred. And I am not immune to moral judgement in everyday life either, to label some of the people I meet selfish, lazy, greedy, aggressive and so on. Often with a muttered swear-word under my breath. I am well aware of the contradiction between this aspect of my own thinking, and what I write in this book. It is another example of the complexities of the mind or brain, a reflection of the difficult nature of what it means to be human. Increasingly, however, in moments of clarity, I find myself asking why people are as they are. Trying to see beneath the surface, to delve into the depths. To grapple with these challenging concepts of the nature and origins of these acts can occasionally provide insights into the 'soul' of an individual, but more importantly informs us what we, as a human society, can do to make the world a little better.

In reality I suspect the answer as to what degree of personal responsibility we hold is somewhere in between. It is likely that, as with all aspects of humanity, there is a spectrum when it comes to the underlying explanation for our behaviours, especially when it comes to our 'sins'. As I have shown, these kinds of behaviours have an evolutionary basis, a point to our existence, and are woven into the thread of our biology. At one extreme are those individuals with devastating injury or disease, in whom those mechanisms that rein in our basic instincts are obliterated, these primal processes underlying these tendencies unleashed. At the other end, are those in whom their genetics, their upbringing, their environment, all unite to constrain these qualities or traits. But in between these two groups of individuals, there is a continuum of those in whom their brains – a function of their genetics and foetal development but also of their experiences – define their behaviours. The unmet challenge is how to define where one single individual sits within this continuum.

It may be that neuroscience, philosophy and even quantum physics give us clearer answers over the next years or decades – maybe. In the meantime, however, to consider someone as intrinsically good or bad is far from straightforward. Can someone's moral worth really be defined by an accident, a drug, a random genetic mutation or an environmental influence on brain development? There are many biological factors that lead a person to their ultimate destination.

I am drawn back to a conversation with Rhett, whose wife, Becky, was so devastated by Huntington's disease. He continues to wrestle in his own mind as to whether her painful words and deeds are a function of her Huntington's or a basic reflection of her mind or soul. At one point he tells me that, despite the clear ravaging of her brain by the disease, he believes that she still knows right from wrong, that she continues to bear some moral responsibility for her actions. Yet it is clear from many individuals that, even when this 'moral compass' remains intact, there are other factors at play. I think of Jono and Tom, and their rage induced by medication and brain injury, who obviously know right from wrong, whose abilities to determine what constitutes correct behaviour remain untarnished. Despite this, in that particular moment those biological processes reining in our intrinsic emotions and associated behaviour simply fail to contain them.

Even the apparently uncomplicated concept of knowing right from wrong is problematic. To do so requires not only an understanding of a moral code, but also a full insight into the implications and consequences for the individual and those around them. On discussion, Rhett acknowledges that it is not at all clear that Becky fully understands the repercussions of her decisions.

The laws of humankind

Beyond these moral aspects, there is the legal angle too. If an individual cannot fully be held responsible for their actions, then does it make a mockery of our criminal justice system? Can we really be viewed as culpable or innocent?

Our legal system to some extent already has mechanisms in place for when a more obvious neurological or psychiatric cause explains a particular act.[17] In the British legal system, for most offences, guilt requires the accused to have performed the *actus reus* (guilty act) and have the *mens rea* (guilty mind). A psychiatrist, neurologist or neuropsychiatrist may be called upon to evaluate whether the defendant had the capacity to form the intent to commit a particular guilty act. The absence of a *mens rea* may represent a legal defence, and this legal principle is of very long standing. The McNaughton rules,* dating back to

---

* Daniel McNaughton (sometimes spelled as MacNaughton or M'Naghten) was found not guilty by reason of insanity after he shot and killed someone he incorrectly believed to be the Prime Minister, Robert Peel.

1843, state: 'To establish a defence on the ground of insanity, it must be clearly proved that, at the time of committing the act, the party accused was labouring under such a defect of reason, from disease of the mind, as to not know the nature and quality of the act he was doing; or, if he did know it, that he did not know what he was doing was wrong.' Note the term 'disease of the mind', which in my view should read 'disease of the brain', although in fairness, these words were written almost two centuries ago.

There is the medico-legal concept of automatism, defined as 'lack of a willed action'. Essentially, any medical condition that results in complete loss of control over one's own actions may fit the bill. Automatisms can be defined as 'an abrupt change in behaviour, in the absence of conscious awareness or memory formation associated with certain specific clinical disorders, such as epilepsy, parasomnia [such as sleepwalking], hypogly-caemia [very low blood sugar levels], and head injury'.[18] As I have written about before in *The Nocturnal Brain*, British law distinguishes 'sane' and 'insane' automatisms. These distinc-tions rest upon whether the medical cause is internal, i.e. emanating from the mind (or brain); or external, such as drugs or alcohol taken involuntarily, psychological or physical trauma; and whether there is a likelihood of it recurring. Like all ques-tions in this area of the law, these issues are very grey, enough to keep judges, lawyers and medico-legal experts busy for many years to come. For example, if a criminal act is as a result of being given insulin, causing the blood sugar to drop precipi-tously low and thus causing neurological dysfunction, this is deemed a sane automatism, as there is an external cause; while if the act was in the context of a diabetic's blood sugar becoming dangerously high (essentially an 'internal' cause), this would be defined as insane. Similarly, a trance-like state resulting from a severe traumatic event, like an accident or rape, would be viewed as sane, but without a clear trigger would be viewed as insane. From a medical perspective, this is somewhat challenging and not entirely sensible.

These legal principles have been developed over hundreds of years, with bases in more primitive understandings of brain and mind. Yet even within this framework, there is an intrinsic acceptance that, for some people at least, medical issues give rise to a lack of legal responsibility. But if medical issues can do so, what about these other factors, like our genetics, or the effects of our upbringing on the structure and function of our brains? Can all of us be argued to have diminished responsibility, or indeed a total lack of responsibility for the decisions that we make, for the crimes and misdemeanours that we commit?

I know that many people will disagree vehemently with the view that any of us have diminished responsibility, except in one or two specific cases. Yet even with this view – of our actions being outside our control – as the starting point, it does not in any way lessen the need for criminal justice. It just nudges the emphasis. Our society still has the need to maintain our safety and our rights. To encourage an environment where certain forms of behaviour are seen as acceptable and unacceptable has an important role for our brains. A simple example: if domestic or sexual violence in childhood has an impact on brain development, and fundamentally predisposes to psychiatric illness or destructive behaviours, then of course we should discourage a society in which these are prevalent. To label an act as being legal or illegal is still appropriate, part of the framework to regulate society and protect us. To imprison someone capable of these harmful acts is no less valid, to protect others from these individuals. And there is the intense need in almost all of us, a natural desire for justice.

But as I said, perhaps this is a question of emphasis. Perhaps the labelling of an individual as good or evil is unhelpful, despite the crimes they commit being labelled as such. Perhaps the question we should be asking ourselves is: 'Is the brain of this individual fixable?' Can we do something to alter the circuitry, to change brain function, to chivvy the neurobiological bases of harmful behaviours towards a better way of life? For some, this will be impossible, and they will need to remain incarcerated. For others, much more achievable, with rehabilitation. While the criminal justice system serves a clear purpose, perhaps we should focus less on the view that it is for punishment of a morally defective criminal, and more about deciding who can be rehabilitated, who cannot, and taking appropriate action.

As I write these words, I admit to being somewhat surprised at myself. I consider myself a progressive in many ways, but have always taken a slightly more reactionary approach to crime and punishment, to morality and guilt. However, there are strands of neuroscience that have reached the same destination as many philosophers like Galen Strawson and Derk Pereboom, albeit from a different perspective: if indeed there is an absence of free will, criminal justice is (or should be) a mechanism for the protection of society and rehabilitation, rather than a meter of morality.

The laws of God

As with the criminal justice system, these themes of individual responsibility are equally applicable when it comes to the moral frameworks placed upon us by religion. They achieve the same through the laws of God, or gods, rather than the laws of the land. Whether it be the Ten Commandments, an explicit list of rules, or Pope Gregory's Seven Deadly Sins – those habits seen as the root cause of all other immoralities – they all seek to guide our conduct in a manner to minimise harm and maximise the contentment of wider society. Rather than punishment by the courts, it is punishment in the Afterlife (or indeed the disapproval of others) that is the stick, as Dante's *Divine Comedy* clearly describes. Beyond simply guiding the individual, their net effect is also to mould the norms of our communities. And as we have seen in the pages of this book, our environment, the culture that we reside in and mature in, has an important influence on our brain development and the neurobiological substrates of our own behaviour. Our upbringing, and what we see around us as normal, is perhaps as important as our genes or anatomy for the functioning of our brains. Examples abound. As described, considering obesity as a communicable disease – that one of the major risk factors for obesity is living with other obese people. In societies where physical violence is prevalent, you are more likely to beat your partner. Where people who are sexually abused in childhood are more likely to become abusers. In times and places, like Nazi Germany, when a normalisation of killing 'the other' resulted in the industrialisation of murder.

As with criminal justice, I am equally surprised by the shift in my views on these theological concepts. As an atheist, I have a natural disinclination to embrace any religion's strictures or rules, and I feel profoundly uncomfortable with the labelling of an individual as good or evil, destined to burn in Hell or ascend to Heaven. But if you take away the religious window-dressing, these moral systems all do the same thing. They are a guide for living one's life in a way that is both for the personal and the greater good. And if one considers a moral act as being something that promotes happiness or reduces harm, then these are a guide for moral behaviours, if not a measuring stick for one's intrinsic moral worth.

* * *

Ultimately, there resides some truth in the Christian doctrine of original sin, that we are all tainted by a sinful nature from the moment of our birth. As per Romans 3:23–4: 'For all have sinned and have fallen in the eyes of God.' The question remains, however, as to what constitutes sin. That these emotions and deeds course through our veins, the very lifeblood of humanity, is irrefutable. What is equally undeniable is that our propensity to these 'sins' is central to our existence, to our survival as a race. Without them, we are destined to extinction. But as with all facets of the human race, it is likely a matter of degree. We all sit somewhere on a spectrum of these tendencies, defined by our biology – our genetics, our evolution, our physical make-up, our brain structure and function. There is no hard border between our innate human natures and the concept of sin though, no fence or signpost to mark the boundaries of normality and evil.

While we may have limited or no personal influence over these aspects of who we are, as a society we can shape the world that we live in, the milieu which we inhabit. As I have shown, our environment can mould our brains, and can sculpt those potent forces within us.

In the containment of our basic instincts, our primal needs and wants, we all need some guidance. These moral or ethical codes – be they religious, philosophical or legal – serve to balance the needs of our societies with those aspects of our biology that sustain the delicate flame of human survival.

Acknowledgements

A book of this type is a different proposition to most. By its nature, it touches upon areas of human experience that are stigmatised, moralised over and riven with shame. Without the individuals described within, there would be no book, and I am eternally grateful to all those people and their families who have entrusted me with their experiences. They have been enormously courageous in their public telling of their stories, and I am both in awe of their fortitude and in their debt. I am also grateful to researcher extraordinaire, Jen Kerrison, for bringing these people together.

As ever, thanks must go to my agent Luigi Bonomi for bringing *Seven Deadly Sins* to fruition. The seeds of this book originated during a relaxed lunch in central London, and with his help and input, as well his colleagues George Lucas and Nicki Kennedy, this project has come to exist. Every time that I say I am not going to write another book, Luigi persuades me otherwise.

I count myself extremely fortunate to have two of the best editors in the business, Arabella Pike at HarperCollins in London, and Michael Flamini at St Martin's Press in New York. Their guidance has been fundamental to making this book a little more intelligible and readable than it otherwise would have been. My friends Anna Davies and Richard Ambrose had the patience to read through earlier drafts of the book, and to provide much-needed feedback to make it better. Thanks too to Kate Johnson for her amazing eye for detail in the copy-editing process, and Eve Hutchings and Sam Harding, who were both instrumental to the production of this book.

My previous two books were written in a furious haze of full-time work. This time, I thought better of trying to repeat this. I was very fortunate to be given a leave of absence by my employer, Guys and St Thomas' NHS Foundation Trust, and I am particularly grateful to Nick Hart for this, as well as to my colleagues who covered my clinical work for this period – special mentions for John Baker, Neil Munro, Laura Perez Carbonell, Elisa Bruno, Fran Watson and Jen Nightingale.

Looking at the very long list of references, it is clear just how much work is going on in the field of neuroscience research. I have relied very heavily on a massive number of researchers around the world who dedicate their lives to understanding the essence of human nature and the brain. As starting points for psychological, theological and philosophical research, I depended on several books, most notably Simon Laham's *The Joy of Sin*, Solomon Schimmel's *The Seven Deadly Sins* and *Free Will* by Michael McKenna and Derk Pereboom. I must also thank my friends and colleagues Al Santhouse and Thomasin Andrews, who provided a keen eye for the psychiatry and neurology within. Any remaining errors or misinterpretations are entirely my own.

On a personal level, I must thank a few people. Firstly, my two goddaughters, Anna Reinaud and Nellie Woods, one of whom is now pursuing a career in medicine, and the other whose interest in neuroscience has been piqued by my books. I am very mindful that my own interest in the wonderful world of neurology was triggered by reading previous iterations of these sorts of popular neuroscience books, and am keen to pay it forward. Also, the Turner family, who have always been a rock of support.

Finally, my own family. My parents, who have instilled in me an intellectual curiosity combined with a rich cultural background. While our household was entirely irreligious, there was always a clear sense of morality, of right and wrong, which has influenced this book and its origins greatly. My daughters, Maya and Ava, have been a source of joy and pride, and have always humoured me as I have ascended to my study to write. Most importantly of all, my wife, Kavita, who at various times has acted as editor, creative director, counsellor, psychologist and confidante. Without her, this book would have been infinitely poorer.

Glossary

amygdala – a small almond-shaped structure located deep in the temporal lobe on either side of the brain. It forms part of the limbic system, and its major role is in the processing of fear and aggression.

anhedonia – the loss of ability to feel pleasure. It is a common symptom in depression and other related psychiatric disorders.

anticipation – the phenomenon of a genetic condition becoming more severe and appearing earlier in age through the generations.

apathy – the medical term for a lack of motivation or goal-directed activity, usually in the context of a neurological disorder.

basal ganglia – a group of nuclei deep within the brain that have a fundamental role in the control of movement, in addition to roles in learning, emotions and executive function. Huntington's and Parkinson's diseases primarily affect these areas of the brain.

Bereitschaftspotential (**BP**) – an electrical signature visible on the EEG before voluntary active movements. It has been proposed as a biological marker of the brain's readiness to act or move.

cognitive dysfunction – deficits in a wide array of higher neurological functions, such as visual and auditory processing, problem-solving, attention and processing speed.

delusions – firmly held beliefs that are evidently false, and are unshakeable, despite clear evidence to the contrary, not accounted for by cultural or religious background or a person's intelligence. Along with hallucinations, these are one of the cardinal features of psychosis.

disinhibition – a tendency towards immediate gratification of urges, without consideration for consequences.

disorder – an abnormality of bodily function, that can be physical, mental, genetic or biochemical. A disorder may relate to a disease, but is a descriptive term of altered function, even in the absence of sufficient evidence to make a specific diagnosis.

electroencephalogram (EEG) – the recording of the electrical activity of the brain, usually through the application of electrodes to the scalp.

epigenetics – the study of how the environment alters the functions of our genes, without altering our genetic code.

executive function – the mental skills that govern our ability to function in daily life, such as working memory, flexible thinking and self-control. These skills are necessary to paying attention, organisational tasks and planning, maintaining focus and regulating our emotions.

frontal lobe – the lobes of the brain lying immediately behind the forehead, concerned with voluntary movement, learning, personality and behaviour.

frontotemporal dementia (FTD) – a group of neurodegenerative disorders that have a predilection for causing the loss of nerve cells in the frontal or temporal lobes. Depending on where the dominant area of cell loss is, it can result in significant behavioural changes or impaired functions of language or meaning.

functional imaging – modalities of medical imaging that aim to visualise function rather than structure of the body. In neuroscience, the most widely used method is functional MRI, which measures small changes in blood flow that occur with brain activity. These changes in blood flow are considered a proxy marker for the underlying metabolic and electrical activity of the brain, and can be used to identify regions that show changes during particular tasks or actions.

genetic mutation – an alteration in the DNA sequence that arises during the division and replication of cells within the body. A mutation may or may not give rise to a disease, depending on the significance of the change, and whether it has sufficient consequences on the structure and function of the product of a particular gene.

genetic variant – if a mutation in the genetic sequence arises, but its effects are not particularly severe or even confer some benefits, it may be passed on. Over generations, this mutation may become more frequent in the population gradually representing a common variant in the DNA sequence rather than a mutation. These common variants, sometimes termed polymorphisms, may have no health effects at all, but can also increase susceptibility to certain diseases, or have health benefits.

glioblastoma multiforme (GBM) – a fast-growing and aggressive tumour arising from astrocytes, cells within the brain substance that support neurones and provide a housekeeping role within the brain. GBM is the most common malignant brain tumour, and untreated, usually results in death within a few months.

grandiosity – a sense of invulnerability, uniqueness or superiority that is not rooted in the truth of one's own personal abilities.

hallucinations – the false perception of sensory experiences, typically visual or auditory. Hallucinations, along with delusions, are the hallmarks of psychosis, but hallucinations can occur in normal individuals or those with neurological disorders. In psychosis, individuals will believe their hallucinations to be real, whereas in other disorders, people will have insight, knowing that their hallucinations do not represent reality.

hypersexuality – excessive or unwanted sexual arousal, causing people to think or engage in sexual activity to the extent that it causes them problems in their life or affects those around them.

hypothalamus – this deep area of the brain is situated behind the eyes. Within it are located several nuclei, responsible for the control of hunger, thirst, sleep, sexual behaviour and our body clocks.

iatrogenic – where an illness is caused by medical intervention or treatment.

imprinting – the genetic phenomenon of silencing a group of genes inherited from one or other parent.

impulse control disorders – a group of behavioural conditions that result from an inability to regulate certain impulses or behaviours to the extent that they are harmful. In the context of neurological disease, they are an adverse effect of drugs used to treat Parkinson's disease, primarily manifesting as compulsive shopping or gambling, overeating or hypersexuality.

leptin – a protein hormone produced by fat cells that functions as a messenger to the brain, conveying body stores of fat. Leptin is one of the major regulators of hunger and satiety.

levodopa – a precursor to dopamine, used as a treatment for Parkinson's disease. When taken orally, levodopa enters the brain, where it is converted to dopamine, normalising levels within the basal ganglia.

limbic system – a collection of brain regions involved in controlling our emotional responses and associated behaviours. The limbic system comprises: the hippocampus, the amygdala, areas of the cerebral cortex, hypothalamus, nucleus accumbens and other brain regions.

neurotransmitters – chemical messengers that the nervous system uses to transmit impulses from one nerve cell to another, across a synapse.

nuclei – clusters of nerve cells with similar connections that serve the same function.

nucleotides – the basic building blocks of nucleic acids such as DNA. Our genetic sequence is defined by how these nucleotides are connected in long chains.

nucleus accumbens – a fundamental brain region in the processing of reward and pleasure, driving motivation.

obsessive-compulsive disorder (OCD) – a mental health condition defined by the presence of obsessions (unwanted or unpleasant thoughts, images or urges that repeatedly occur, causing anxiety or unease) and compulsions (repetitive acts, either mental or physical, that momentarily relieve the unpleasant feelings associated with obsessive thoughts).

paraphilias – intense or recurrent sexual arousal to atypical situations, objects, behaviours or individuals. What 'atypical' means is debatable, since the definition of unconventional sexual interests shifts between cultures and over time.

personality disorder – a way of acting, feeling or thinking that deviates from what is viewed as normal within the cultural or religious background causing distress or functioning in life. Personality disorders represent a magnification of normal personality traits that are sufficient to cause problems. They usually begin in adolescence or early adulthood.

pre-frontal cortex (PFC) – the surface of the frontal lobes closest to the forehead. This brain region is responsible for our personalities, social behaviour, planning, decision-making and ability to suppress urges.

psychopathy – a personality disorder characterised by antisocial behaviour, deficiencies in remorse and emotional empathy, and egotistical traits.

psychosis – a collection of symptoms characterised by a loss of contact with reality, typically through delusions and hallucinations. There is an associated loss of insight, in that people with psychosis believe their hallucinations to be real; whereas hallucinations with insight into their true nature, i.e. an awareness that these hallucinations are not real, can occur in individuals with specific neurological disorders, or even in healthy people.

sectioning – legal detainment in a medical facility for psychiatric reasons under the Mental Health Act 1983. The duration of initial detainment can range from six hours to six months, but can be extended.

supplementary motor area (SMA) – a region of the motor cortex that contributes to the control of movement. It also appears to have a crucial role in feedback of physical exertion to sensory parts of the brain.

temporal lobe – the lobe of the brain that sits behind the ears. Its major functions are in language, memory and deriving meaning from verbal, visual or emotional cues.

ventromedial pre-frontal cortex (vmPFC) – part of the pre-frontal cortex that is most implicated in the assessment of our emotions and actions related to them; decision-making, self-control and social interaction.

Figures

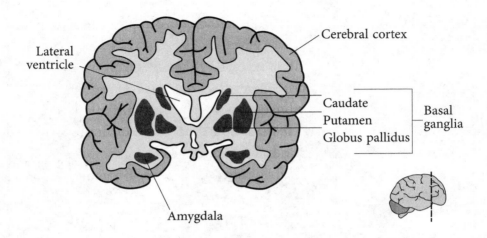

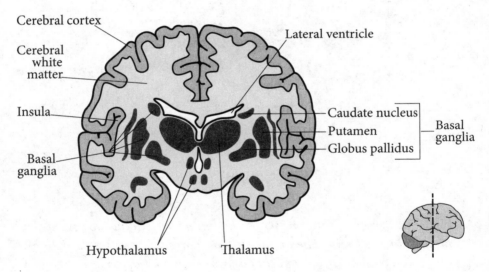

Figure 1. *Slices through the brain, parallel to the face. The basal ganglia are crucial for movement and motivation, and are involved in conditions like Parkinson's disease and Huntington's disease. A hallmark of Huntington's is the withering of the caudate nucleus, with resultant apparent enlargement of the fluid-filled lateral ventricles. The amygdala is the almond-shaped structure deep in the temporal lobe, and is a key component in fear and emotional processing. The hypothalamus contains multiple specialised areas r:esponsible for the regulation of functions and behaviours critical to survival (see Figure 4).*

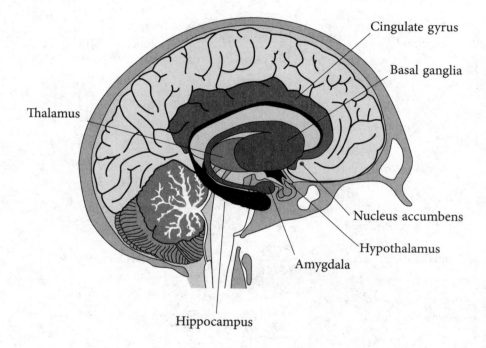

Figure 2. *Primary components of the limbic system, viewed from the midline of the brain. The limbic system represents a network of brain areas that support basic emotional processing and behaviours, as well as memory and smell. The hippocampus and amygdala are located deep within the temporal lobe, offset from the midline (see Figure 1).*

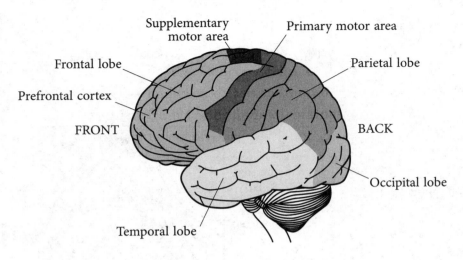

Figure 3A. (**Above**) *External surface of brain, viewed from the side. The frontal lobe is the largest lobe of the brain. The rearmost part of the frontal lobe is largely dedicated to movement, but most of it comprises the prefrontal cortex (PFC). The supplementary motor area (SMA) is implicated in the monitoring of physical effort. Disruption of neural activity in the SMA using magnetic pulses can lessen physical fatigue.* **Figure 3B.** (**Below**) *Locations of the ventromedial PFC, most intimately linked to areas of the brain involved in emotions, pleasure and social behaviour, and the dorsolateral PFC, which regulates so-called 'executive functions' – planning, abstract reasoning and working memory.*

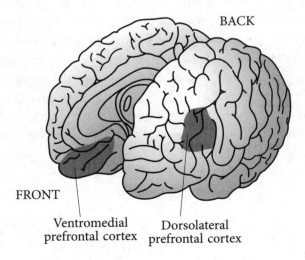

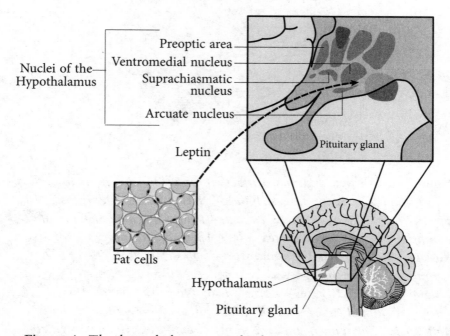

Figure 4. *The hypothalamus. Multiple tiny nuclei have specialised functions to support those behaviours crucial to survival. These include: the detection of leptin, the chemical signal produced by fat cells as a marker of fat stores and a regulator of appetite, by the arcuate nucleus, the ventromedial hypothalamus and other nuclei; the sexually dimorphic nucleus within the preoptic area, which differs hugely in size between males and females; the suprachiasmatic nucleus, where our circadian clock is encoded; the pre-optic area, which coordinates sleep and body temperature; and other areas that regulate hormone secretion, thirst, and regulation of blood pressure and sexual behaviour.*

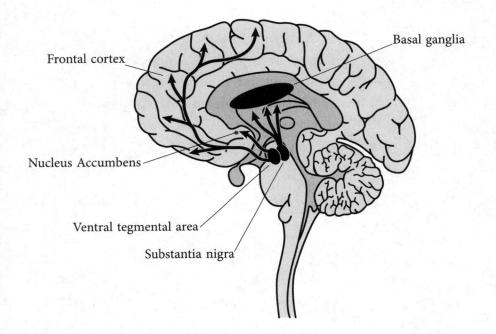

Figure 5. *Circuits within the brain relying primarily on dopamine as a chemical transmitter. One circuit originates in the substantia nigra and projects to the basal ganglia. Loss of dopamine-producing nerve cells in the substantia nigra causes Parkinson's disease, and its movement abnormalities. The mesolimbic pathway, projects from the ventral tegmental area to the nucleus accumbens, 'the pleasure centre', and is a key component of reward and pleasure, implicated in addiction and regulation of eating, sex and other survival behaviours.*

Notes

Chapter 1. Wrath

1. Kanemoto, K., Tadokoro, Y. and Oshima, T. Violence and postictal psychosis: A comparison of postictal psychosis, interictal psychosis, and postictal confusion. *Epilepsy Behav.* 19, 162–166 (2010).
2. Singh, R. *et al.* Characteristics and Neural Correlates of Emotional Behavior during Prefrontal Seizures. *Ann. Neurol.* 92, 1052–1065 (2022).
3. Pottkämper, J. C. M., Hofmeijer, J., Waarde, J. A. van and Putten, M. J. A. M. van. The postictal state – What do we know? *Epilepsia* 61, 1045 (2020).
4. Mikulincer, M. Reactance and helplessness following exposure to unsolvable problems: the effects of attributional style. *J. Pers. Soc. Psychol.* 54, 679–686 (1988).
5. Author interview with Jono and Hannah (June 2023).
6. Chen, Z. *et al.* Psychotic disorders induced by antiepileptic drugs in people with epilepsy. *Brain* 139, 2668–2678 (2016).
7. Zhang, J. F., Piryani, R., Swayampakula, A. K. and Farooq, O. Levetiracetam-induced aggression and acute behavioral changes: A case report and literature review. *Clin. Case Rep.* 10, e05586 (2022).
8. Wilkowski, B. M. and Robinson, M. D. The Cognitive Basis of Trait Anger and Reactive Aggression: An Integrative Analysis. *Personal. Soc. Psychol. Rev.* 12, 3–21 (2008).
9. Lievaart, M., van der Veen, F. M., Huijding, J., Hovens, J. E. and Franken, I. H. A. The Relation Between Trait Anger and Impulse Control in

Forensic Psychiatric Patients: An EEG Study. *Appl. Psychophysiol. Biofeedback* 43, 131–142 (2018).

10. Scott, S. K. *et al.* Impaired auditory recognition of fear and anger following bilateral amygdala lesions. *Nature* 385, 254–257 (1997).

11. Richard, Y., Tazi, N., Frydecka, D., Hamid, M. S. and Moustafa, A. A. A systematic review of neural, cognitive, and clinical studies of anger and aggression. *Curr. Psychol.* (2022) doi:10.1007/s12144-022-03143-6.

12. Leschziner, G. *The Nocturnal Brain.* Simon and Schuster (2019).

13. Teffer, K. and Semendeferi, K. Human prefrontal cortex, in *Progress in Brain Research.* vol. 195. 191–218 (Elsevier, 2012).

14. Hathaway, W. R. and Newton, B. W. Neuroanatomy, Prefrontal Cortex, in *StatPearls* (StatPearls Publishing, Treasure Island (FL), 2023).

15. Klimecki, O. M., Sander, D. and Vuilleumier, P. Distinct Brain Areas involved in Anger versus Punishment during Social Interactions. *Sci. Rep.* 8, 10556 (2018).

16. Richard *et al.* A systematic review . . . *Curr. Psychol.* (2022).

17. World Health Organization. Third milestones of a Global Campaign for Violence Prevention report, 2007: scaling up. 31 (2007).

18. Björkqvist, K. Gender differences in aggression. *Curr. Opin. Psychol.* 19, 39–42 (2018).

19. Sarkar, A. and Wrangham, R. W. Evolutionary and neuroendocrine foundations of human aggression. *Trends Cogn. Sci.* 27, 468–493 (2023).

20. Booth, A. and Dabbs, J. M. Testosterone and Men's Marriages. *Soc. Forces* 72, 463–477 (1993).

21. McIntyre, M. H. *et al.* Finger length ratio (2D:4D) and sex differences in aggression during a simulated war game. *Personal. Individ. Differ.* 42, 755–764 (2007).

22. Millet, K. and Dewitte, S. Digit ratio (2D:4D) moderates the impact of an aggressive music video on aggression. *Personal. Individ. Differ.* 43, 289–294 (2007).

23. Bailey, A. A. and Hurd, P. L. Finger length ratio (2D:4D) correlates with physical aggression in men but not in women. *Biol. Psychol.* 68, 215–222 (2005).

24. Björkqvist, Gender differences . . . *Curr. Opin. Psychol.* (2018).

25. Siever, L. J. Neurobiology of aggression and violence. *Am. J. Psychiatry* 165, 429–442 (2008).

26. Brunner, H. G., Nelen, M., Breakefield, X. O., Ropers, H. H. and van

Oost, B. A. Abnormal behavior associated with a point mutation in the structural gene for monoamine oxidase A. *Science* 262, 578–580 (1993).

27. Godar, S. C., Fite, P. J., McFarlin, K. M. and Bortolato, M. The role of monoamine oxidase A in aggression: Current translational developments and future challenges. *Prog. Neuropsychopharmacol. Biol. Psychiatry* 69, 90–100 (2016).

28. McSwiggan, S., Elger, B. and Appelbaum, P. S. The forensic use of behavioral genetics in criminal proceedings: Case of the MAOA-L genotype. *Int. J. Law Psychiatry* 50, 17–23 (2017).

29. Cupaioli, F. A. *et al.* The neurobiology of human aggressive behavior: Neuroimaging, genetic, and neurochemical aspects. *Prog. Neuropsychopharmacol. Biol. Psychiatry* 106, 110059 (2021).

30. Leichsenring, F. *et al.* Borderline Personality Disorder: A Review. *JAMA* 329, 670–679 (2023).

31. Newhill, C. E., Eack, S. M. and Mulvey, E. P. Violent behavior in borderline personality. *J. Personal. Disord.* 23, 541–554 (2009).

32. Hengartner, M. P., Ajdacic-Gross, V., Rodgers, S., Müller, M. and Rössler, W. Childhood adversity in association with personality disorder dimensions: New findings in an old debate. *Eur. Psychiatry* 28, 476–482 (2013).

33. Cattane, N., Rossi, R., Lanfredi, M. and Cattaneo, A. Borderline personality disorder and childhood trauma: exploring the affected biological systems and mechanisms. *BMC Psychiatry* 17, 221 (2017).

34. Cattane *et al.* Borderline personality disorder . . . *BMC Psychiatry* (2017).

35. Dammann, G. *et al.* Increased DNA methylation of neuropsychiatric genes occurs in borderline personality disorder. *Epigenetics* 6, 1454–1462 (2011).

36. Cupaioli *et al.* The neurobiology of human aggressive behavior . . . *Prog. Neuropsychopharmacol. Biol. Psychiatry* (2021).

37. Caspi, A. *et al.* Role of Genotype in the Cycle of Violence in Maltreated Children. *Science* 297, 851–854 (2002).

38. Barnes, J. C., Beaver, K. M. and Boutwell, B. B. A Functional Polymorphism in a Serotonin Transporter Gene (5-HTTLPR) Interacts with 9/11 to Predict Gun-Carrying Behavior. *PLOS ONE* 8, e70807 (2013).

39. Siever, Neurobiology of aggression and violence. *Am. J. Psychiatry* (2008).

40. Author interview with Tom and Han (May 2023).

41. Williams, W. H. *et al.* Traumatic brain injury: a potential cause of violent crime? *Lancet Psychiatry* 5, 836–844 (2018).

42. Max, J. E. *et al.* Predictors of personality change due to traumatic brain injury in children and adolescents in the first six months after injury. *J. Am. Acad. Child Adolesc. Psychiatry* 44, 434–442 (2005).

43. Stoddard, S. A. and Zimmerman, M. A. Association of interpersonal violence with self-reported history of head injury. *Pediatrics* 127, 1074–1079 (2011).

44. Grafman, J. *et al.* Frontal lobe injuries, violence, and aggression: a report of the Vietnam Head Injury Study. *Neurology* 46, 1231–1238 (1996).

45. Williams *et al.* Traumatic brain injury. *Lancet Psychiatry* (2018).

46. Fazel, S., Lichtenstein, P., Grann, M. and Långström, N. Risk of Violent Crime in Individuals with Epilepsy and Traumatic Brain Injury: A 35-Year Swedish Population Study. *PLoS Med.* 8, e1001150 (2011).

47. Schiltz, K., Witzel, J. G., Bausch-Hölterhoff, J. and Bogerts, B. High prevalence of brain pathology in violent prisoners: a qualitative CT and MRI scan study. *Eur. Arch. Psychiatry Clin. Neurosci.* 263, 607–616 (2013).

48. Parsonage, M. *Traumatic Brain Injury and Offending: An Economic Analysis.* Centre for Mental Health (2016).

49. Landberg, J. and Norström, T. Alcohol and homicide in Russia and the United States: a comparative analysis. *J. Stud. Alcohol Drugs* 72, 723–730 (2011).

50. Sontate, K. V. *et al.* Alcohol, Aggression, and Violence: From Public Health to Neuroscience. *Front. Psychol.* 12 (2021).

51. Wrangham, R. W. Two types of aggression in human evolution. *Proc. Natl. Acad. Sci.* 115, 245–253 (2018).

52. Sarkar and Wrangham, Evolutionary and neuroendocrine . . . *Trends Cogn. Sci.* (2023).

53. De Brito, S. A. *et al.* Psychopathy. *Nat. Rev. Dis. Primer* 7, 1–21 (2021).

54. Johanson, M., Vaurio, O., Tiihonen, J. and Lähteenvuo, M. A Systematic Literature Review of Neuroimaging of Psychopathic Traits. *Front. Psychiatry* 10, 1027 (2020).

55. Grossman, D. *On Killing: The Psychological Cost of Learning to Kill in War and Society.* Little, Brown (2009).

Chapter 2. Gluttony

1. Yanagihara, H. *To Paradise*. Picador (2022).

2. Ringel, M. M. and Ditto, P. H. The moralization of obesity. *Soc. Sci. Med.* 237, 112399 (2019).

3. Rigano, K. S. *et al.* Life in the fat lane: seasonal regulation of insulin sensitivity, food intake, and adipose biology in brown bears. *J. Comp. Physiol. B* 187, 649–676 (2017).

4. Author interview with Alex (June 2023).

5. Rahman, Q. F. Ab., Jufri, N. F. and Hamid, A. Hyperphagia in Prader-Willi syndrome with obesity: From development to pharmacological treatment. *Intractable Rare Dis. Res.* 12, 5–12 (2023).

6. Author interview with Kate and Jon (June 2023).

7. Hetherington, A. W. and Ranson, S. W. The Spontaneous Activity and Food Intake of Rats with Hypothalamic Lesions. *Am. J. Physiol.-Leg. Content* 136, 609–617 (1942).

8. Hervey, G. R. The effects of lesions in the hypothalamus in parabiotic rats. *J. Physiol.* 145, 336-352.3 (1959).

9. Hervey, The effects of lesions . . . *J. Physiol.* (1959).

10. Ingalls, A. M., Dickie, M. M. and Snell, G. D. Obese, a new mutation in the house mouse. *J. Hered.* 41, 317–318 (1950).

11. Barsh, G. S., Farooqi, I. S. and O'Rahilly, S. Genetics of body-weight regulation. *Nature* 404, 644–651 (2000).

12. Ranadive, S. A. and Vaisse, C. Lessons from Extreme Human Obesity: Monogenic Disorders. *Endocrinol. Metab. Clin. North Am.* 37, 733–x (2008).

13. Ranadive and Vaisse, Lessons from Extreme Human Obesity . . . *Metab. Clin. North Am.* (2008).

14. Locke, A. E. *et al.* Genetic studies of body mass index yield new insights for obesity biology. *Nature* 518, 197–206 (2015).

15. Neel, J. V. Diabetes Mellitus: A "Thrifty" Genotype Rendered Detrimental by "Progress"? *Am. J. Hum. Genet.* 14, 353–362 (1962).

16. Wu, T. and Xu, S. Understanding the contemporary high obesity rate from an evolutionary genetic perspective. *Hereditas* 160, 5 (2023).

17. Speakman, J. R. A nonadaptive scenario explaining the genetic predisposition to obesity: the 'predation release' hypothesis. *Cell Metab.* 6, 5–12 (2007).

18. Sellayah, D., Cagampang, F. R. and Cox, R. D. On the evolutionary

origins of obesity: a new hypothesis. *Endocrinology* 155, 1573–1588 (2014).

19. Jia, J. *et al.* The polymorphisms of UCP1 genes associated with fat metabolism, obesity and diabetes. *Mol. Biol. Rep.* 37, 1513–1522 (2010).

20. Moore, T. and Haig, D. Genomic imprinting in mammalian development: a parental tug-of-war. *Trends Genet.* 7, 45–49 (1991).

21. Ma, V. K. *et al.* Prader-Willi and Angelman Syndromes: Mechanisms and Management. *Appl. Clin. Genet.* 16, 41–52 (2023).

22. Wang, Y. *et al.* Melanocortin 4 receptor signals at the neuronal primary cilium to control food intake and body weight. *J. Clin. Invest.* 131, e142064.

23. Holsen, L. M. *et al.* Neural mechanisms underlying hyperphagia in Prader-Willi syndrome. *Obes. Silver Spring Md* 14, 1028–1037 (2006).

24. Price, R. A. and Gottesman, I. I. Body fat in identical twins reared apart: roles for genes and environment. *Behav. Genet.* 21, 1–7 (1991).

25. Christakis, N. A. and Fowler, J. H. The Spread of Obesity in a Large Social Network over 32 Years. *N. Engl. J. Med.* 357, 370–379 (2007).

26. Atkinson, R. L. Viruses as an Etiology of Obesity. *Mayo Clin. Proc.* 82, 1192–1198 (2007).

27. Dhurandhar, N. V., Kulkarni, P. R., Ajinkya, S. M., Sherikar, A. A. and Atkinson, R. L. Association of Adenovirus Infection with Human Obesity. *Obes. Res.* 5, 464–469 (1997).

28. Atkinson, R. L. *et al.* Human adenovirus-36 is associated with increased body weight and paradoxical reduction of serum lipids. *Int. J. Obes.* 29, 281–286 (2005).

29. Vangipuram, S. D. *et al.* Adipogenic human adenovirus-36 reduces leptin expression and secretion and increases glucose uptake by fat cells. *Int. J. Obes.* 31, 87–96 (2007).

30. Kirk, R. G. W. 'Life in a Germ-Free World': Isolating Life from the Laboratory Animal to the Bubble Boy. *Bull. Hist. Med.* 86, 237–275 (2012).

31. Ley, R. E., Turnbaugh, P. J., Klein, S. and Gordon, J. I. Human gut microbes associated with obesity. *Nature* 444, 1022–1023 (2006).

32. Hasan, N. and Yang, H. Factors affecting the composition of the gut microbiota, and its modulation. *PeerJ* 7, e7502 (2019).

33. Zhang, F., Luo, W., Shi, Y., Fan, Z. and Ji, G. Should We Standardize the 1,700-Year-Old Fecal Microbiota Transplantation? *Off. J. Am. Coll. Gastroenterol. ACG* 107, 1755 (2012).

34. Hasan and Yang. Factors affecting the composition . . . *PeerJ* (2019).

35. Bäckhed, F. *et al.* The gut microbiota as an environmental factor that regulates fat storage. *Proc. Natl. Acad. Sci. U. S. A.* 101, 15718–15723 (2004).

36. Cox, L. M. *et al.* Altering the intestinal microbiota during a critical developmental window has lasting metabolic consequences. *Cell* 158, 705–721 (2014).

37. Van Hul, M. and Cani, P. D. The gut microbiota in obesity and weight management: microbes as friends or foe? *Nat. Rev. Endocrinol.* 1–14 (2023) doi:10.1038/s41574-022-00794-0.

38. Van Hul and Cani. The gut microbiota in obesity and weight management . . . *Nat. Rev. Endocrinol.* (2023).

39. Roseboom, T., de Rooij, S. and Painter, R. The Dutch famine and its long-term consequences for adult health. *Early Hum. Dev.* 82, 485–491 (2006).

40. Roseboom, Rooji and Painter. The Dutch famine . . . *Early Hum. Dev.* (2006).

41. Godfrey, K. M. *et al.* Influence of maternal obesity on the long-term health of offspring. *Lancet Diabetes Endocrinol.* 5, 53–64 (2017).

42. Gaillard, R., Steegers, E. a. P., Franco, O. H., Hofman, A. and Jaddoe, V. W. V. Maternal weight gain in different periods of pregnancy and childhood cardio-metabolic outcomes. The Generation R Study. *Int. J. Obes. 2005* 39, 677–685 (2015).

43. Samuelsson, A.-M. *et al.* Diet-induced obesity in female mice leads to offspring hyperphagia, adiposity, hypertension, and insulin resistance: a novel murine model of developmental programming. *Hypertens. Dallas Tex 1979* 51, 383–392 (2008).

44. Taylor, P. D., Samuelsson, A.-M. and Poston, L. Maternal obesity and the developmental programming of hypertension: a role for leptin. *Acta Physiol. Oxf. Engl.* 210, 508–523 (2014).

45. Şanlı, E. and Kabaran, S. Maternal Obesity, Maternal Overnutrition and Fetal Programming: Effects of Epigenetic Mechanisms on the Development of Metabolic Disorders. *Curr. Genomics* 20, 419–427 (2019).

46. Zhang, J., Li, S., Luo, X. and Zhang, C. Emerging role of hypothalamus in the metabolic regulation in the offspring of maternal obesity. *Front. Nutr.* 10, 1094616 (2023).

47. Horsthemke, B. A critical view on transgenerational epigenetic inheritance in humans. *Nat. Commun.* 9, 2973 (2018).

48. Volkow, N. D., Wise, R. A. and Baler, R. The dopamine motive system: implications for drug and food addiction. *Nat. Rev. Neurosci.* 18, 741–752 (2017).

49. Speranza, L., di Porzio, U., Viggiano, D., de Donato, A. and Volpicelli, F. Dopamine: The Neuromodulator of Long-Term Synaptic Plasticity, Reward and Movement Control. *Cells* 10, 735 (2021).

50. Wise, R. A. Catecholamine theories of reward: a critical review. *Brain Res.* 152, 215–247 (1978).

51. Montague, P. R., Dayan, P. and Sejnowski, T. J. A framework for mesencephalic dopamine systems based on predictive Hebbian learning. *J. Neurosci. Off. J. Soc. Neurosci.* 16, 1936–1947 (1996).

52. Small, D. M., Jones-Gotman, M. and Dagher, A. Feeding-induced dopamine release in dorsal striatum correlates with meal pleasantness ratings in healthy human volunteers. *NeuroImage* 19, 1709–1715 (2003).

53. Volkow, N. D., Wang, G.-J. and Baler, R. D. Reward, dopamine and the control of food intake: implications for obesity. *Trends Cogn. Sci.* 15, 37–46 (2011).

54. Yamamoto, R. T., Foulds-Mathes, W. and Kanarek, R. B. Antinociceptive actions of peripheral glucose administration. *Pharmacol. Biochem. Behav.* 117, 34–39 (2014).

55. Avena, N. M., Rada, P. and Hoebel, B. G. Evidence for sugar addiction: Behavioral and neurochemical effects of intermittent, excessive sugar intake. *Neurosci. Biobehav. Rev.* 32, 20–39 (2008).

56. Frayn, M. and Knäuper, B. Emotional Eating and Weight in Adults: a Review. *Curr. Psychol.* 37, 924–933 (2018).

57. Wang, G.-J. *et al.* Brain dopamine and obesity. *The Lancet* 357, 354–357 (2001).

58. Volkow, Wise and Baler. The dopamine motive system . . . *Nat. Rev. Neurosci.* (2017).

59. Guillaumin, M. C. C. and Peleg-Raibstein, D. Maternal Over- and Malnutrition and Increased Risk for Addictive and Eating Disorders in the Offspring. *Nutrients* 15, 1095 (2023).

60. Izquierdo, A. G., Crujeiras, A. B., Casanueva, F. F. and Carreira, M. C. Leptin, Obesity, and Leptin Resistance: Where Are We 25 Years Later? *Nutrients* 11, 2704 (2019).

61. Rahman, Jufri and Hamid. Hyperphagia in Prader-Willi syndrome . . . *Intractable Rare Dis. Res.* (2023).

Chapter 3. Lust

1. Jarvie, H. F. Frontal Lobe Wounds causing Disinhibition. *J. Neurol. Neurosurg. Psychiatry* 17, 14–32 (1954).
2. Author interview with Simon (May 2023).
3. Freud, S. *The Standard Edition of the Complete Psychological Works of Sigmund Freud.* Macmillan (1964).
4. Winch, R. F. *Mate-Selection; a Study of Complementary Needs.* xix, 349. Harper (1958).
5. Clark, M. S. and Reis, H. T. Interpersonal Processes in Close Relationships. *Ann. Rev. Psychol.* 39: 609–72 (1988).
6. Horwitz, T. B., Balbona, J. V., Paulich, K. N. and Keller, M. C. Evidence of correlations between human partners based on systematic reviews and meta-analyses of 22 traits and UK Biobank analysis of 133 traits. *Nat. Hum. Behav.* 1–16 (2023) doi:10.1038/s41562-023-01672-z.
7. Stefka, J. *et al.* Misattributed parentage identified through diagnostic exome sequencing: Frequency of detection and reporting practices. *J. Genet. Couns.* 31, 631–640 (2022).
8. Voracek, M., Haubner, T. and Fisher, M. L. Recent Decline in Nonpaternity Rates: A Cross-Temporal Meta-Analysis. *Psychol. Rep.* 103, 799–811 (2008).
9. Buss, D. M. and Schmitt, D. P. Sexual strategies theory: an evolutionary perspective on human mating. *Psychol. Rev.* 100, 204–232 (1993).
10. Buss, D. M. and Schmitt, D. P. Mate Preferences and Their Behavioral Manifestations. *Annu. Rev. Psychol.* 70, 77–110 (2019).
11. Schmitt, D. P. Sociosexuality from Argentina to Zimbabwe: a 48-nation study of sex, culture, and strategies of human mating. *Behav. Brain Sci.* 28, 247–275 (2005).
12. Lippa, R. A. Sex differences in sex drive, sociosexuality, and height across 53 nations: testing evolutionary and social structural theories. *Arch. Sex. Behav.* 38, 631–651 (2009).
13. Li, N. P. Mate Preference Necessities in Long- and Short-Term Mating: People Prioritize in Themselves What Their Mates Prioritize in Them. *Acta Psychol. Sin.* 39, 528–535 (2007).
14. Valentine, K. A., Li, N. P., Penke, L. and Perrett, D. I. Judging a Man by the Width of his Face: The Role of Facial Ratios and Dominance in Mate Choice at Speed-Dating Events. *Psychol. Sci.* 25, 806–811 (2014).
15. Muggleton, N. K. and Fincher, C. L. Unrestricted sexuality promotes

distinctive short- and long-term mate preferences in women. *Personal. Individ. Differ.* 111, 169–173 (2017).

16. Gildersleeve, K., Haselton, M. G. and Fales, M. R. Do women's mate preferences change across the ovulatory cycle? A meta-analytic review. *Psychol. Bull.* 140, 1205–1259 (2014).

17. Glass, S. P. and Wright, T. L. Justifications for extramarital relationships: The association between attitudes, behaviors, and gender. *J. Sex Res.* 29, 361–387 (1992).

18. Buss, D. M. Sex differences in human mate preferences: Evolutionary hypotheses tested in 37 cultures. *Behav. Brain Sci.* 12, 1–49 (1989).

19. Wang, G. *et al.* Different impacts of resources on opposite sex ratings of physical attractiveness by males and females. *Evol. Hum. Behav.* 39, 220–225 (2018).

20. Marlowe, F. W. Mate preferences among Hadza hunter-gatherers. *Hum. Nat.* 15, 365–376 (2004).

21. Pawlowski, B. and Koziel, S. The impact of traits offered in personal advertisements on response rates. *Evol. Hum. Behav.* 23, 139–149 (2002).

22. Sugiyama, L. S. Physical Attractiveness in Adaptationist Perspective, in *The Handbook of Evolutionary Psychology* 292–343. John Wiley & Sons (2005).

23. Sohn, K. Men's revealed preference for their mates' ages. *Evol. Hum. Behav.* 38, 58–62 (2017).

24. Buss and Schmitt, Mate Preferences . . . *Annu. Rev. Psychol.* (2019).

25. Buss, D. *The Evolution of Desire: Strategies of Human Mating*. Basic Books (2016).

26. Miller, B. L., Cummings, J. L., McIntyre, H., Ebers, G. and Grode, M. Hypersexuality or altered sexual preference following brain injury. *J. Neurol. Neurosurg. Psychiatry* 49, 867–873 (1986).

27. Miller, Cummings *et al.* Hypersexuality . . . *J. Neurol. Neurosurg. Psychiatry* 49, 867–873 (1986).

28. Surbeck, W., Bouthillier, A. and Nguyen, D. K. Bilateral cortical representation of orgasmic ecstasy localized by depth electrodes. *Epilepsy Behav. Case Rep.* 1, 62–65 (2013).

29. Miller, Cummings *et al.* Hypersexuality . . . *J. Neurol. Neurosurg. Psychiatry* 49, 867–873 (1986).

30. Henn, F. A., Herjanic, M. and Vanderpearl, R. H. Forensic psychiatry:

profiles of two types of sex offenders. *Am. J. Psychiatry* 133, 694–696 (1976).

31. Swaab, D. F. and Garcia-Falgueras, A. Sexual differentiation of the human brain in relation to gender identity and sexual orientation. *Functional Neurology*, 24: 17–28 (2009).

32. Phoenix, C. H., Goy, R. W., Gerall, A. A. and Young, W. C. Organizing Action of Prenatally Administered Testosterone Propionate on the Tissues Mediating Mating Behavior in the Female Guinea Pig. *Endocrinology* 65, 369–382 (1959).

33. Dessens, A. B. *et al*. Prenatal exposure to anticonvulsants and psychosexual development. *Arch. Sex. Behav.* 28, 31–44 (1999).

34. Health Check: The boy who was raised a girl. www.bbc.co.uk/news/health-11814300. BBC News. (23 Nov. 2010).

35. Health Check: The boy who was raised a girl.

36. Colapinto, J. *As Nature Made Him: The Boy Who Was Raised as a Girl.* Harper Perennial (2006).

37. Swaab and Garcia-Falgueras, A. Sexual differentiation of the human brain . . . *Functional Neurology* (2009).

38. Stock, K. *Material Girls: Why Reality Matters for Feminism.* Fleet (2021).

39. Kinnunen, L. H., Moltz, H., Metz, J. and Cooper, M. Differential brain activation in exclusively homosexual and heterosexual men produced by the selective serotonin reuptake inhibitor, fluoxetine. *Brain Res.* 1024, 251–254 (2004).

40. Savic, I. Brain Imaging Studies of the Functional Organization of Human Olfaction. *Chem. Senses* 30, i222–i223 (2005).

41. Paul, T. *et al*. Brain response to visual sexual stimuli in heterosexual and homosexual males. *Hum. Brain Mapp.* 29, 726–735 (2007).

42. Savic, I. and Lindström, P. PET and MRI show differences in cerebral asymmetry and functional connectivity between homo- and heterosexual subjects. *Proc. Natl. Acad. Sci. U. S. A.* 105, 9403–9408 (2008).

43. Miller, Cummings *et al*. Hypersexuality . . . *J. Neurol. Neurosurg. Psychiatry* 49, 867–873 (1986).

44. Barbeau, A. L-Dopa Therapy in Parkinson's Disease. *Can. Med. Assoc. J.* 101, 59–68 (1969).

45. Cummings, J. L. Behavioral Complications of Drug Treatment of Parkinson's Disease. *J. Am. Geriatr. Soc.* 39, 708–716 (1991).

46. Solla, P., Bortolato, M., Cannas, A., Mulas, C. S. and Marrosu, F. Paraphilias and paraphilic disorders in Parkinson's disease: A systematic review of the literature. *Mov. Disord.* 30, 604–613 (2015).

47. Nielssen, O. B., Cook, R. J., Joffe, R., Meagher, L. J. and Silberstein, P. Paraphilia and other disturbed behavior associated with dopamimetic treatment for Parkinson's disease. *Mov. Disord. Off. J. Mov. Disord. Soc.* 24, 1091–1092 (2009).

48. Raina, G., Cersosimo, M. G. and Micheli, F. Zoophilia and impulse control disorder in a patient with Parkinson disease. *J. Neurol.* 259, 969–970 (2012).

49. Solla, P., Bortolato, M., Cannas, A., Mulas, C. S. and Marrosu, F. Paraphilias and paraphilic disorders in Parkinson's disease: A systematic review of the literature. *Mov. Disord.* 30, 604–613 (2015).

50. Coombs, R. H. *Handbook of Addictive Disorders: A Practical Guide to Diagnosis and Treatment.* John Wiley & Sons (2004).

51. Ahmad, Z. S. *et al.* Prevalence Rates of Online Sexual Addiction Among Christian Clergy. *Sex. Addict. Compulsivity* 22, 344–356 (2015).

52. Toates, F. A motivation model of sex addiction – Relevance to the controversy over the concept. *Neurosci. Biobehav. Rev.* 142, 104872 (2022).

53. Potenza, M. N., Gola, M., Voon, V., Kor, A. and Kraus, S. W. Is excessive sexual behaviour an addictive disorder? *Lancet Psychiatry* 4, 663–664 (2017).

54. Black, D. W. Compulsive Sexual Behavior: A Review. *J. Psychiatr. Pract.* 4, 219 (1998).

55. Toates. A motivation model of sex addiction. *Neurosci. Biobehav. Rev.* (2022).

56. Gold, S. N. and Heffner, C. L. Sexual addiction: Many conceptions, minimal data. *Clin. Psychol. Rev.* 18, 367–381 (1998).

57. Voon, V. *et al.* Neural Correlates of Sexual Cue Reactivity in Individuals with and without Compulsive Sexual Behaviours. *PLoS ONE* 9, e102419 (2014).

58. Gold and Heffner, Sexual addiction. *Clin. Psychol.* (1998).

59. Author interview with Simon.

60. Hayley, S., Vahid-Ansari, F., Sun, H. and Albert, P. R. Mood disturbances in Parkinson's disease: From prodromal origins to application of animal models. *Neurobiol. Dis.* 181, 106115 (2023).

Chapter 4. Envy

1. Author interview with Sarah and Colin (June 2023).

2. Ramachandran, V. S. and Jalal, B. The Evolutionary Psychology of Envy and Jealousy. *Front. Psychol.* 8, 1619 (2017).

3. Aquaro, G. *Death by Envy: The Evil Eye and Envy in the Christian Tradition.* iUniverse, Inc. (2004).

4. Rand, A., Schwartz (ed. and contributor). *Return of the Primitive: The Anti-Industrial Revolution.* Penguin. (1999).

5. Smith, R. H. and Kim, S. H. Comprehending envy. *Psychol. Bull.* 133, 46 (20070103).

6. Laham, S. *The Joy of Sin: The Psychology of the Seven Deadly Sins.* Constable (2012).

7. Ramachandran and Jalal. The Evolutionary Psychology . . . *Front. Psychol.* (2017).

8. Chen-Charash, Y. and Larson, E. What Is the Nature of Envy? in: *Envy at Work and in Organizations.* Oxford University Press (2017).

9. Frances, A. DSM-5 Is a Guide, Not a Bible: Simply Ignore Its 10 Worst Changes. *HuffPost.* www.huffpost.com/entry/dsm-5_b_2227626 (2012).

10. Newlin, E. and Weinstein, B. Personality Disorders. *Contin. Lifelong Learn. Neurol.* 21, 806 (2015).

11. Berenz, E. C. *et al.* Childhood Trauma and Personality Disorder Criterion Counts: A Co-twin Control Analysis. *J. Abnorm. Psychol.* 122, 1070–1076 (2013).

12. Torgersen, S. *et al.* A twin study of personality disorders. *Compr. Psychiatry* 41, 416–425 (2000).

13. Newlin and Weinstein. Personality Disorders. *Contin. Lifelong Learn. Neurol.* (2015).

14. Kendell, R. E. The distinction between personality disorder and mental illness. *Br. J. Psychiatry* 180, 110–115 (2002).

15. Neufeld D. C. and Johnson, E. A. Burning with envy? Dispositional and situational influences on envy in grandiose and vulnerable narcissism. *J. Pers.* 84: 685–696 (2016).

16. Neufeld and Johnson. Burning with envy? *J. Pers.* (2016).

17. Sanderson, D. Gems thief died after killing wife 'in Othello rage'. *The Times* (30 Aug. 2013).

18. Robson, S. Diamond thief suffering from 'Othello syndrome' strangled

his wife then hanged himself because he believed she was having an affair. *Mail Online.* www.dailymail.co.uk/news/article-2406631/Othello-Syndrome-sufferer-Robert-Mercati-killed-wife-self-delusional-jealousy. html (2013).

19. Shakespeare, W. *Othello*, Act III, sc.iv.

20. Mathes, E. W. Men's desire for children carrying their genes and sexual jealousy: a test of paternity uncertainty as an explanation of male sexual jealousy. *Psychol. Rep.* 96, 791–798 (2005).

21. Edlund, J. E. *et al.* Male Sexual Jealousy: Lost Paternity Opportunities? *Psychol. Rep.* 122, 575–592 (2019).

22. Kingham, M. and Gordon, H. Aspects of morbid jealousy. *Adv. Psychiatr. Treat.* 10, 207–215 (2004).

23. Mooney, H. B. Pathologic Jealousy and Psychochemotherapy. *Br. J. Psychiatry* 111, 1023–1042 (1965).

24. Todd, J., Mackie, J. R. M. and Dewhurst, K. Real or imaginary hypophallism: A cause of inferiority feelings and morbid sexual jealousy. *Brit. J. Psychiatry* 119: 315–318 (1971).

25. Askevold, F. Predictive value of neuropathic traits. *Acta Psychiatr. Scand.* 56, 32–38 (1977).

26. Sun, Y. *et al.* Neural substrates and behavioral profiles of romantic jealousy and its temporal dynamics. *Sci. Rep.* 6, 27469 (2016).

27. Kuruppuarachchi, K. A. L. A. and Seneviratne, A. N. Organic causation of morbid jealousy. *Asian J. Psychiatry* 4, 258–260 (2011).

28. Soyka, M., Naber, G. and Völcker, A. Prevalence of delusional jealousy in different psychiatric disorders. An analysis of 93 cases. *Br. J. Psychiatry J. Ment. Sci.* 158, 549–553 (1991).

29. Alzheimer, A., Stelzmann, R. A., Schnitzlein, H. N. and Murtagh, F. R. An English translation of Alzheimer's 1907 paper, 'Uber eine eigenartige Erkankung der Hirnrinde'. *Clin. Anat. N. Y. N* 8, 429–431 (1995).

30. Hashimoto, M., Sakamoto, S. and Ikeda, M. Clinical Features of Delusional Jealousy in Elderly Patients With Dementia. *J. Clin. Psychiatry* 76, 2769 (2015).

31. Leutmezer, F. *et al.* Postictal Psychosis in Temporal Lobe Epilepsy. *Epilepsia* 44, 582–590 (2003).

32. Steffen, P. R., Hedges, D. and Matheson, R. The Brain is Adaptive Not Triune: How the Brain Responds to Threat, Challenge, and Change. *Front. Psychiatry* 13 (2022).

Chapter 5. Sloth

1. Leschziner, G. *The Nocturnal Brain*. Simon & Schuster (2019).

2. Hicks, A.-I. and Prager-Khoutorsky, M. Neuronal culprits of sickness behaviours. *Nature* 609, 679–680 (2022).

3. Ilanges, A. *et al*. Brainstem ADCYAP1+ neurons control multiple aspects of sickness behaviour. *Nature* 609, 761–771 (2022).

4. Schimmel, S. *The Seven Deadly Sins: Jewish, Christian, and Classical Reflections on Human Nature*. Free Press (1992).

5. Husain, M. and Roiser, J. P. Neuroscience of apathy and anhedonia: a transdiagnostic approach. *Nat. Rev. Neurosci.* 19, 470–484 (2018).

6. Hartmann, M. N. *et al*. Apathy in schizophrenia as a deficit in the generation of options for action. *J. Abnorm. Psychol.* 124, 309–318 (2015).

7. Muhammed, K. *et al*. Reward sensitivity deficits modulated by dopamine are associated with apathy in Parkinson's disease. *Brain* 139, 2706–2721 (2016).

8. Husain and Roiser. Neuroscience of apathy and anhedonia . . . *Nat. Rev. Neurosci.* (2018).

9. Besser, L. M. and Galvin, J. E. Diagnostic experience reported by caregivers of patients with frontotemporal degeneration. *Neurol. Clin. Pract.* 10, 298–306 (2020).

10. Boeve, B. F. Behavioral Variant Frontotemporal Dementia. *Contin. Lifelong Learn. Neurol.* 28, 702 (2022).

11. Shaw, S. R. *et al*. Uncovering the prevalence and neural substrates of anhedonia in frontotemporal dementia. *Brain* 144, 1551–1564 (2021).

12. Levy, R. and Dubois, B. Apathy and the Functional Anatomy of the Prefrontal Cortex–Basal Ganglia Circuits. *Cereb. Cortex* 16, 916–928 (2006).

13. Author interviews with Rhett. (June and Sept. 2023).

14. Bhattacharyya, K. B. The story of George Huntington and his disease. *Ann Indian Acad Neurol.* 19: 25–28 (2016).

15. Huntington, G. On Chorea. *Med Surg Rep.* 26, 317–321 (1872).

16. Walker, F. O. Huntington's disease. *Lancet Lond. Engl.* 369, 218–228 (2007).

17. McColgan, P. and Tabrizi, S. J. Huntington's disease: a clinical review. *Eur. J. Neurol.* 25, 24–34 (2018).

18. Tabrizi, S. J. *et al*. Potential disease-modifying therapies for Huntington's

disease: lessons learned and future opportunities. *Lancet Neurol.* 21, 645–658 (2022).

19. Matmati, J., Verny, C. and Allain, P. Apathy and Huntington's Disease: A Literature Review Based on PRISMA. *J Neuropsychiatry Clin Neurosci.* 34: 100–112 (2022).

20. Tabrizi, S. J. *et al.* Predictors of phenotypic progression and disease onset in premanifest and early-stage Huntington's disease in the TRACK-HD study: analysis of 36-month observational data. *Lancet Neurol.* 12, 637–649 (2013).

21. Levy and Dubois. Apathy . . . *Cereb. Cortex* (2006).

22. Levy and Dubois. Apathy . . . *Cereb. Cortex* (2006).

23. Author interview with AJ. (May 2023).

24. Pérez-Carbonell, L., Mignot, E., Leschziner, G. and Dauvilliers, Y. Understanding and approaching excessive daytime sleepiness. *The Lancet* 400, 1033–1046 (2022).

25. Valko, V. O. and Baumann C. R. Chapter 108: Sleep Disorders after Traumatic Brain Injury, in *Principles and Practice of Sleep Medicine (7th Ed).* Eds Kryger, M., Roth, T., Goldstein, C. A., Dement, W. C. Elsevier (2022).

26. Marcora, S. Perception of effort during exercise is independent of afferent feedback from skeletal muscles, heart, and lungs. *J. Appl. Physiol.* 106, 2060–2062 (2009).

27. Ainley, V., Apps, M. A. J., Fotopoulou, A. and Tsakiris, M. 'Bodily precision': a predictive coding account of individual differences in interoceptive accuracy. *Philos. Trans. R. Soc. B Biol. Sci.* 371, 20160003 (2016).

28. Zénon, A., Sidibé, M. and Olivier, E. Disrupting the Supplementary Motor Area Makes Physical Effort Appear Less Effortful. *J. Neurosci.* 35, 8737–8744 (2015).

29. Boksem, M. A. S., Meijman, T. F. and Lorist, M. M. Mental fatigue, motivation and action monitoring. *Biol. Psychol.* 72, 123–132 (2006).

30. Blain, B., Hollard, G. and Pessiglione, M. Neural mechanisms underlying the impact of daylong cognitive work on economic decisions. *Proc. Natl. Acad. Sci.* 113, 6967–6972 (2016).

31. Müller, T. and Apps, M. A. J. Motivational fatigue: A neurocognitive framework for the impact of effortful exertion on subsequent motivation. *Neuropsychologia* 123, 141–151 (2019).

32. Sloan, M. *et al.* Prevalence and identification of neuropsychiatric symptoms in systemic autoimmune rheumatic diseases: an international mixed

methods study. *Rheumatol. Oxf. Engl.* kead369 (2023) doi:10.1093/rheumatology/kead369.

33. Le Heron, C., Holroyd, C. B., Salamone, J. and Husain, M. Brain mechanisms underlying apathy. *J. Neurol. Neurosurg. Psychiatry* 90, 302–312 (2019).

34. Turkheimer, F. E., Veronese, M., Mondelli, V., Cash, D. and Pariante, C. M. Sickness behaviour and depression: An updated model of peripheral-central immunity interactions. *Brain. Behav. Immun.* 111, 202–210 (2023).

35. Ziino, C. and Ponsford, J. Selective attention deficits and subjective fatigue following traumatic brain injury. *Neuropsychology* 20, 383–390 (2006).

36. Cantor, J. B. *et al.* Fatigue after traumatic brain injury and its impact on participation and quality of life. *J. Head Trauma Rehabil.* 23, 41–51 (2008).

37. Mavroudis, I. *et al.* Functional Overlay Model of Persistent Post-Concussion Syndrome. *Brain Sci.* 13, 1028 (2023).

Chapter 6. Greed

1. Levy, M. 'Interested in data?': Panama Papers leak began with message from 'John Doe'. *Sydney Morning Herald.* www.smh.com.au/business/banking-and-finance/interested-in-data-panama-papers-leak-began-with-message-from-john-doe-20160406-gnzd0i.html (2016).

2. Obama, B. Remarks on Tax Code Reform and an Exchange With Reporters. www.govinfo.gov/content/pkg/DCPD-201600222/pdf/DCPD-201600222.pdf (5 Apr. 2016).

3. Panama Papers: Former Pakistan PM Sharif Sentenced To 10 Years – ICIJ. www.icij.org/investigations/panama-papers/former-pakistan-pm-sharif-sentenced-to-10-years-over-panama-papers/ (2018).

4. *Wall Street.* (20th Century Fox, 1987).

5. Seuntjens, T. G., Zeelenberg, M., van de Ven, N. and Breugelmans, S. M. Greedy bastards: Testing the relationship between wanting more and unethical behavior. *Personal. Individ. Differ.* 138, 147–156 (2019).

6. Seuntjens, T. G., Zeelenberg, M., van de Ven, N. and Breugelmans, S. M. Dispositional greed. *J. Pers. Soc. Psychol.* 108, 917–933 (2015).

7. Zeelenberg, M. and Breugelmans, S. M. The good, bad and ugly of dispositional greed. *Curr. Opin. Psychol.* 46, 101323 (2022).

8. Seuntjens, Zeelenberg and Breugelmans. Greedy bastards . . . *Personal. Individ. Differ.* (2019).

9. Hoyer, K., Zeelenberg, M., Breugelmans, S. M. Greed: What Is It Good for? *Pers. Soc. Psychol. Bull.* (2022) Dec 28:1461672221140355. doi: 10.1177/01461672221140355.

10. Krekels, G. and Pandelaere, M. Dispositional greed. *Personal. Individ. Differ.* 74, 225–230 (2015).

11. Zeelenberg, M., Seuntjens, T. G., van de Ven, N. and Breugelmans, S. M. When enough is not enough: Overearning as a manifestation of dispositional greed. *Personal. Individ. Differ.* 165, 110155 (2020).

12. Zhu, Y., Sun, X., Liu, S. and Xue, G. Is Greed a Double-Edged Sword? The Roles of the Need for Social Status and Perceived Distributive Justice in the Relationship Between Greed and Job Performance. *Front. Psychol.* 10, 2021 (2019).

13. Laham, S. *The Joy of Sin: The Psychology of the Seven Deadly Sins.* Constable (2012).

14. Vohs, K. D., Mead, N. L. and Goode, M. R. The Psychological Consequences of Money. *Science* 314, 1154–1156 (2006).

15. Zhou, X., Vohs, K. D. and Baumeister, R. F. The symbolic power of money: Reminders of money alter social distress and physical pain. *Psychol. Sci.* 20, 700–706 (2009).

16. Hass, R. G. Perspective taking and self-awareness: Drawing an E on your forehead. *J. Pers. Soc. Psychol.* 46, 788–798 (1984).

17. Laham. *The Joy of Sin* (2012).

18. Stevenson, B. and Wolfers, J. Economic Growth and Subjective Well-Being: Reassessing the Easterlin Paradox. Working Paper at: doi. org/10.3386/w14282 (2008).

19. Hoyer, Zeelenberg and Breugelmans. Greed . . . *Pers. Soc. Psychol. Bull.* (2022).

20. Hoyer, Zeelenberg and Breugelmans. Greed . . . *Pers. Soc. Psychol. Bull.* (2022).

21. Bao, R., Sun, X., Liu, Z., Fu, Z. and Xue, G. Dispositional greed inhibits prosocial behaviors: an emotive – social cognitive dual-process model. *Curr. Psychol.* 41, 3928–3936 (2022).

22. Wei, S. *et al.* Greed personality trait links to negative psychopathology and underlying neural substrates. *Soc. Cogn. Affect. Neurosci.* 18, nsac046 (2022).

23. Osnos, E. Life After White-Collar Crime. *The New Yorker* (23 Aug. 2021).

24. Seuntjens, Zeelenberg, and Breugelmans. Dispositional greed. *J. Pers. Soc. Psychol.* (2015).

25. Zeelenberg and Breugelmans. The good, bad and ugly . . . *Curr. Opin. Psychol.* (2022).

26. Zeelenberg and Breugelmans. The good, bad and ugly . . . *Curr. Opin. Psychol.* (2022).

27. Grisham, J. R. and Norberg, M. M. Compulsive hoarding: current controversies and new directions. *Dialogues Clin. Neurosci.* 12, 233–240 (2010).

28. O'Sullivan, S. S. *et al.* Excessive hoarding in Parkinson's disease. *Mov. Disord.* 25, 1026–1033 (2010).

29. Miltner, W. H. R., Braun, C. H. and Coles, M. G. H. Event-Related Brain Potentials Following Incorrect Feedback in a Time-Estimation Task: Evidence for a "Generic" Neural System for Error Detection. *J. Cogn. Neurosci.* 9, 788–798 (1997).

30. Mussel, P. and Hewig, J. A neural perspective on when and why trait greed comes at the expense of others. *Sci. Rep.* 9, 10985 (2019).

31. Mussel, P., Reiter, A. M. F., Osinsky, R. and Hewig, J. State- and trait-greed, its impact on risky decision-making and underlying neural mechanisms. *Soc. Neurosci.* 10, 126–134 (2015).

32. Dikman, Z. V. and Allen, J. J. Error monitoring during reward and avoidance learning in high- and low-socialized individuals. *Psychophysiology* 37, 43–54 (2000).

33. Mussel and Hewig. A neural perspective . . . *Sci. Rep.* 9, 10985 (2019).

34. Wei *et al.* Greed personality trait . . . *Soc. Cogn. Affect. Neurosci.* (2022).

35. Coffey, C. *et al.* Time to Care: Unpaid and underpaid care work and the global inequality crisis. hdl.handle.net/10546/620928 (2020) doi:10.21201/2020.5419.

36. Lambie, G. W. and Haugen, J. S. Understanding greed as a unified construct. *Personal. Individ. Differ.* 141, 31–39 (2019).

Chapter 7. Pride

1. Author interviews with Chris and Wendy (July 2023).

2. Email from Chris (Aug. 2023).

3. Author interview with Chris (July 2023).

4. Larkin, P. This Be The Verse (1972).

5. Kirzan, Z., and Herlache, A. D. The Narcissism Spectrum Model: A Synthetic View of Narcissistic Personality. *Pers. Soc. Psychol. Rev.* 22: 3–31 (2018).

6. Brummelman, E. *et al.* Origins of narcissism in children. *Proc. Natl. Acad. Sci.* 112, 3659–3662 (2015).

7. Brummelman, E., Thomaes, S. and Sedikides, C. Separating Narcissism From Self-Esteem. *Curr. Dir. Psychol. Sci.* 25: 8–13 (2016).

8. Brummelman, E. and Sedikides, C. Raising Children With High Self-Esteem (But Not Narcissism). *Child Dev. Perspect.* 14, 83–89 (2020).

9. Trzesniewski, K. H., Donnellan, M. B. and Robins, R. W. Is Generation Me Really More Narcissistic Than Previous Generations? *J. Pers.* 76, 903 (2008).

10. Twenge, J. M., Konrath, S., Foster, J. D., Campbell, W. K. and Bushman, B. J. Egos inflating over time: a cross-temporal meta-analysis of the Narcissistic Personality Inventory. *J. Pers.* 76, 875–902 (2008).

11. Foster, J. D., Keith Campbell, W. and Twenge, J. M. Individual differences in narcissism: Inflated self-views across the lifespan and around the world. *J. Res. Personal.* 37, 469–486 (2003).

12. Brummelman *et al.* Origins . . . *Proc. Natl. Acad. Sci.* (2015).

13. Brummelman *et al.* Origins . . . *Proc. Natl. Acad. Sci.* (2015).

14. Brummelman and Sedikides. Raising Children . . . *Child Dev. Perspect.* (2020).

15. Aristotle. *Rhetoric.*

16. Weiner, B. *Human Motivation.* Springer (1985). doi:10.1007/978-1-4612-5092-0.

17. Herrald, M. M. and Tomaka, J. Patterns of emotion-specific appraisal, coping, and cardiovascular reactivity during an ongoing emotional episode. *J. Pers. Soc. Psychol.* 83, 434 (20020731).

18. Williams, L. A. and DeSteno, D. Pride and perseverance: The motivational role of pride. *J. Pers. Soc. Psychol.* 94, 1007–1017 (2008).

19. Gnambs, T. and Appel, M. Narcissism and Social Networking Behavior: A Meta-Analysis. *J. Pers.* 86, 200–212 (2018).

20. Owen, D. and Davidson, J. Hubris syndrome: An acquired personality disorder? A study of US Presidents and UK Prime Ministers over the last 100 years. *Brain* 132, 1396–1406 (2009).

21. Osnos, E. Is Political Hubris an Illness? *The New Yorker* (5 May 2017).

22. Owen and Davidson. Hubris syndrome . . . *Brain* (2009).

23. Stinson, F. S. *et al.* Prevalence, Correlates, Disability, and Comorbidity of DSM-IV Narcissistic Personality Disorder: Results from the Wave 2 National Epidemiologic Survey on Alcohol and Related Conditions. *J. Clin. Psychiatry* 69, 1033–1045 (2008).

24. Ronningstam, E. Narcissistic personality disorder: a clinical perspective. *J. Psychiatr. Pract.* 17, 89–99 (2011).

25. Caligor, E., Levy, K. N. and Yeomans, F. E. Narcissistic personality disorder: diagnostic and clinical challenges. *Am. J. Psychiatry* 172, 415–422 (2015).

26. Jauk, E. and Kanske, P. Can neuroscience help to understand narcissism? A systematic review of an emerging field. *Personal. Neurosci.* 4, e3 (2021).

27. di Giacomo, E., Andreini, E., Lorusso, O. and Clerici, M. The dark side of empathy in narcissistic personality disorder. *Front. Psychiatry* 14, 1074558 (2023).

28. Lamm, C., Decety, J. and Singer, T. Meta-analytic evidence for common and distinct neural networks associated with directly experienced pain and empathy for pain. *NeuroImage* 54, 2492–2502 (2011).

29. Fan, Y. *et al.* The narcissistic self and its psychological and neural correlates: an exploratory fMRI study. *Psychol. Med.* 41, 1641–1650 (2011).

30. Schulze, L. *et al.* Gray matter abnormalities in patients with narcissistic personality disorder. *J. Psychiatr. Res.* 47, 1363–1369 (2013).

31. Karl Jaspers. *General Psychopathology.* Springer (1913).

32. Isham, L. *et al.* Understanding, treating, and renaming grandiose delusions: A qualitative study. *Psychol. Psychother.* 94, 119–140 (2021).

33. Isham, L. *et al.* The meaning in grandiose delusions: measure development and cohort studies in clinical psychosis and non-clinical general population groups in the UK and Ireland. *Lancet Psychiatry* 9, 792–803 (2022).

34. Isham, L. *et al.* The Difficulties of Grandiose Delusions: Harms, Challenges, and Implications for Treatment Engagement. *Schizophr. Bull.* sbad016 (2023) doi:10.1093/schbul/sbad016.

35. Leschziner, G. *The Man Who Tasted Words.* Simon and Schuster/St Martin's Press (2022).

36. Coltheart, M. The neuropsychology of delusions. *Ann. N. Y. Acad. Sci.* 1191, 16–26 (2010).

37. Corlett, P. R. *et al*. Disrupted prediction-error signal in psychosis: evidence for an associative account of delusions. *Brain J. Neurol*. 130, 2387–2400 (2007).

38. Darby, R. R., Laganiere, S., Pascual-Leone, A., Prasad, S. and Fox, M. D. Finding the imposter: brain connectivity of lesions causing delusional misidentifications. *Brain J. Neurol*. 140, 497–507 (2017).

39. Garety, P. A. *et al*. Differences in Cognitive and Emotional Processes Between Persecutory and Grandiose Delusions. *Schizophr. Bull*. 39, 629–639 (2013).

40. Sabri, O. *et al*. Correlation of positive symptoms exclusively to hyperperfusion or hypoperfusion of cerebral cortex in never-treated schizophrenics. *Lancet Lond. Engl*. 349, 1735–1739 (1997).

41. Kelley, W. M. *et al*. Finding the self? An event-related fMRI study. *J. Cogn. Neurosci*. 14, 785–794 (2002).

42. Kimhy, D., Goetz, R., Yale, S., Corcoran, C. and Malaspina, D. Delusions in individuals with schizophrenia: factor structure, clinical correlates, and putative neurobiology. *Psychopathology* 38, 338–344 (2005).

43. Email from Chris. (Nov. 2023).

44. Cummings, J. L. Organic Delusions: Phenomenology, Anatomical Correlations, and Review. *Brit. J. Psychiatry*. 146: 184–197 (1985).

45. Kermani, E., Drob, S. and Alpert, M. Organic brain syndrome in three cases of acquired immune deficiency syndrome. *Compr. Psychiatry* 25, 294–297 (1984).

46. Sultan, S. and Omar Fallata, E. A Case of Complex Partial Seizures Presenting as Acute and Transient Psychotic Disorder. *Case Rep. Psychiatry* 2019, 1901254 (2019).

47. Dubovsky, A. N., Arvikar, S., Stern, T. A. and Axelrod, L. The Neuropsychiatric Complications of Glucocorticoid Use: Steroid Psychosis Revisited. *Psychosomatics* 53, 103–115 (2012).

48. Manea, M. M. *et al*. Can a Manic Be Organic? *Eur. Psychiatry* 30, 1925 (2015).

49. Benrimoh, D. *et al*. Why We Still Use "Organic Causes": Results From a Survey of Psychiatrists and Residents. *J. Neuropsychiatry Clin. Neurosci*. 31, 57–64 (2019).

50. Das, A. and Khanna, R. Organic manic syndrome: causative factors, phenomenology and immediate outcome. *J. Affect. Disord*. 27, 147–153 (1993).

Chapter 8. Free Will

1. McKenna, M., and Pereboom, D. *Free Will: A Contemporary Introduction*. Routledge and CRC (2016).

2. Watson, G. 4: Responsibility and the Limits of Evil: Variations on a Strawsonian Theme, in *Perspectives on Moral Responsibility*. Eds. Fischer, J. M. and Ravizza, M. 119–148. Cornell University Press (1987).

3. Faure, P. From accouchement to agony: a lexicological analysis of words of French origin in the modern English language of medicine. *Lexis J. Engl. Lexicology* (2018) doi:10.4000/lexis.1171.

4. Colebatch, J. G. Bereitschaftspotential and movement-related potentials: Origin, significance, and application in disorders of human movement. *Mov. Disord.* 22, 601–610 (2007).

5. Libet, B., Gleason, C. A., Wright, E. W. and Pearl, D. K. Time of conscious intention to act in relation to onset of cerebral activity (readiness-potential). The unconscious initiation of a freely voluntary act. *Brain J. Neurol.* 106 (Pt 3), 623–642 (1983).

6. Libet *et al.* Time of conscious intention . . . *Brain J. Neurol.* (1983).

7. Neafsey, E. J. Conscious intention and human action: Review of the rise and fall of the readiness potential and Libet's clock. *Conscious. Cogn.* 94, 103171 (2021).

8. Jung, R. Voluntary intention and conscious selection in complex learned action. *Behav. Brain Sci.* 8, 544–545 (1985).

9. Neafsey, Conscious intention . . . *Conscious. Cogn.* (2021).

10. Schurger, A., Sitt, J. D. and Dehaene, S. An accumulator model for spontaneous neural activity prior to self-initiated movement. *Proc. Natl. Acad. Sci.* 109, E2904–E2913 (2012).

11. Neafsey, Conscious intention . . . *Conscious. Cogn.* (2021).

12. Fried, I., Mukamel, R. and Kreiman, G. Internally generated preactivation of single neurons in human medial frontal cortex predicts volition. *Neuron* 69, 548–562 (2011).

13. Soon, C. S., Brass, M., Heinze, H.-J. and Haynes, J.-D. Unconscious determinants of free decisions in the human brain. *Nat. Neurosci.* 11, 543–545 (2008).

14. Stone, J., Hoeritzauer, I., McWhirter, L. and Carson, A. Functional neurological disorder: defying dualism. *World Psychiatry Off. J. World Psychiatr. Assoc. WPA* 23, 53–54 (2024).

15. Libet, B. Do We Have Free Will? *J. Conscious. Stud.* 6, 47–57 (1999).

16. Spinoza, B. *Ethics.* (1677).

17. Eastman, N., Poole, N. A. and Kopelman, M. D. Neuropsychiatry in the criminal courts, in *Oxford Textbook of Neuropsychiatry.* Eds. Agrawal, N. *et al.* Oxford University Press (2020).

18. Kopelman, M. Memory Disorders in the Law Courts. *Medico-legal J.* 81: 18–28 (2013).

Index

About the Author

DR. GUY LESCHZINER is the author of *The Nocturnal Brain* and *The Man Who Tasted Words,* as well as a professor of neurology and sleep medicine in London. He sees patients with a range of neurological and sleep disorders, and is actively involved in research and teaching. He has also presented series on sleep and neurology for BBC World Service and Radio 4.